PRAISE FOR *THE PROFITABLE FARM*

"Agriculture has been waiting for this book for at least a century. The MSO model brings clarity to the process of primary food production; energy conversion is at its heart and it cannot be divorced from sound farm business management. The implications of the book are profound and of significance worldwide."

– Andrew Hattan, Dairy Farmer and Cheesemaker

"For the first time since 1975 UK farmers are facing the very real prospect of farming without government support. With 80% of farms not currently viable without subsidy now perhaps is the time to re-evaluate your farm business, *The Profitable Farm* provides the first, fully farmer-focused financial assessment mechanism which will make any farmer who embraces the journey to MSO completely re-evaluate the practical and financial management of their farm."

– Andrew Jamieson - Jamieson and Jamieson Ltd

"*The Profitable Farm* shows how to keep control of your costs and inputs to make a business that will be here tomorrow."

– Tim Parton, Farm Manager, Brewood Park Farm
& Innovation Farmer of the Year (2020)

"Nature and profitability are often seen to be in conflict with each other, yet they are closely entwined. Finally, in *The Profitable Farm* comes solid evidence for how they are dependent on each other, why sustainable farming is essential and why farming doesn't follow the conventional rules previously established. The concepts in this book combat the current economic instability and biodiversity crisis we are facing as a nation."

– Tara Wright, 4th Generation Exmoor Farmer

"A truly thoughtful contribution to the potential future for such a critical aspect of the economy."

– Andrew Marsden MStJ, Corporate Strategy Consultant

"Ignoring the fundamental thermodynamic laws of physics leads us to economic miscalculation and needlessly lower profitability in this most noble profession of growing human sustaining energy – or food as we usually call it."
– Debbie Trebilco PhD FRSC, Scientist and Smallholder

"This book encourages farmers and advisors to drill down into their business accounts in a structured way and plan a way forward that improves outcomes for their family, their bank balance and the environment. An important task for us all in these increasingly challenging times."
– Professor Julia Aglionby, University of Cumbria & Commissioner, the Food, Farming & Countryside Commission

"The authors brilliantly pour detailed farm business data, natural assets and landscapes, plus energy and nutrient cycles all into a black box. The emerging calculations – combined with their expertise to deliver a profitable and sustainable approach and based on Maximum Sustainable Output – are genius. Recommended for all farmers, policymakers and nature advocates."
– Vicki Hird MSc FRES, Strategic Lead on Agriculture at The Wildlife Trusts

"This book encourages farmers and advisors to drill down into their business accounts in a structured way and plan a way forward that improves outcomes for their family, their bank balance and the environment."
– Professor Julia Aglionby, University of Cumbria & Commissioner, the Food, Farming & Countryside Commission

"For those of us trying to work out how to make farming profitable this is important reading. The authors understand that there is a sweet spot in farm management at which profits are maximised and that is not necessarily achieved by maximising outputs. As support payments become more vulnerable, this is an important lesson to learn."
– David Fursdon, Dyson Farming Chair

THE PROFITABLE FARM

THE
PROFITABLE
FARM

BALANCING BUSINESS, NATURE AND
ENERGY THROUGH MAXIMUM
SUSTAINABLVE OUTPUT

Chris Clark & Brian Scanlon

 5m Books

First published 2025, revised reprint 2025

Published by
5m Books Ltd
Lings, Great Easton
Essex CM6 2HH, UK
Tel: +44 (0)330 1333 580
www.5mbooks.com

A Catalogue record for this book is available from the British Library

ISBN 9781917159098
eISBN 9781917159203
DOI 10.52517/9781917159203

EU GPSR Authorised Representative
LOGOS EUROPE, 9 rue Nicolas Poussin, 17000, LA ROCHELLE, France
E-mail: Contact@logoseurope.eu

Book layout by Toynbee Editorial Services Ltd
Printed by Hobbs the Printers
Photos and illustrations by the authors unless otherwise indicated
Cover illustration: shutterstock_2460830153

This book is dedicated to the memory of Glye Hodson. Our great friend and mentor, famous for his aphorisms. Two, of many: "Banks don't lend you money, they lend you time" and "You can't eat a percentage."

CONTENTS

FOREWORD

I first met Chris Clark through our work with the Nature Friendly Farming Network. As our conversations turned to Maximum Sustainable Output (MSO) – a financial analysis that struck me as both refreshingly straightforward and incredibly powerful – I was introduced to Brian Scanlon. Like many farmers and land managers, I have always sought ways to balance financial stability with the long-term health of my land. MSO provided a clear, data-driven approach to achieving that balance.

Applying its principles to my own business, I saw firsthand how shifting towards MSO made both financial and environmental sense. One of the most significant changes we made was incorporating more grass into our rotation – a decision that continues to improve soil health and profitability. MSO isn't about cutting costs at the expense of sustainability or driving production to unsustainable levels. Instead, it helps farmers optimise output by making the most of free natural resources, such as sunlight and water, ensuring long-term resilience.

That's why this book is so valuable. Chris and Brian lay out the principles of MSO in a way that is accessible, and – most importantly – pragmatic. Whether you're a farmer, land manager, or anyone responsible for making decisions about land and business sustainability, this book offers a roadmap to financial resilience and environmental stewardship working in harmony.

If you're looking for a way to secure the future of your business while improving the land you manage, Maximum Sustainable Output is a concept worth understanding – and this book is the perfect place to start.

Johnnie Balfour CA
Managing Partner, Balbirnie Home Farms
Chair, Pasture for Life
February 2025

PREFACE

In the short time since the book was published, in April 2025, there has been a large mailbag of comments and reviews from across the farming community, which deserve a considered response through the introduction of this preface and some additional material to Chapter 10.

The comments from the farming community have been invariably positive but often indicate the need for a simple, single statement about the underlying concepts of maximum sustainable output (MSO). The comments from the academic community highlight a specific issue. Standard economic theory treats variable cost functions as **continuous** whereas MSO theory is founded on the phenomenon that variable costs in farming have **sequential** components.

In essence, the MSO proposition is that:

> When the landscape is farmed so that it harnesses all the available natural resources on a farm to produce food, without the interventions of any artificial substitutes for Nature or industrial accelerants, there will be a common point on every farm where its profitability is maximised, its energy footprint is minimised, and its natural capital is on a path of increasing improvement to its maximum potential. This is its MSO point and Nature will exhibit a unique characteristic pattern, on the farm, that will be optimised for commercial food production. This pattern, which will differ from farm to farm, will have one common feature – Nature is always uncompromised at the MSO point.

The discontinuous (sequential) nature of variable costs in farming is a direct consequence of there being two very different forms of energy available to farmers.

- Solar energy can be treated on a free-issue basis inferring that the sun represents an infinite source of energy where the effects of decay (entropy) can be ignored in the big picture.

- However, industrial energy embedded in artificial inputs cannot be treated in this way and conversion losses cannot be ignored. Economic theories currently have no recognition of entropy related effects and the authors have been 'assured' by some economists that these effects are irrelevant. Their perspective needs to change.

At MSO:

- farm profitability is maximised
- nature is optimised
- the energy burden is minimised
- natural capital is maximised.

The concept of MSO also has implications for national food security policies. The great mistake made in the farming sector of the economy has been the belief that the pursuit of improvements in yield cannot be anything other than beneficial. Policy makers and economists have consistently pushed this mantra, yet the profession has failed to recognise that farming is part of a wider energy sector. Driving output with artificial energy-intensive enhancers is the road to ultimate commercial ruin and environmental breakdown. Maximising output should not be the objective; maximising farm profitability by working at MSO is a better policy.

Farming is a highly fragmented economic sector and its fortunes are often determined by the large, remote corporations that comprise the top of the supply-chain pyramid. These corporations have some common features. They all have professional purchasing departments, which often conclude that fewer larger suppliers would reduce their transaction costs. This works against the farming community. They are all now concerned with continuity of supply (in volume and quality terms) but rarely elevate this above the pursuit of price advantage. There are now some anxieties that this is pushing farming into an untenable position and perhaps this is a signal that attitudes are about to change.

All this has led to a growing interest in the concept of resilience and some new material has been added to Chapter 10 on this subject.

The broad scope of this book was a conscious decision and it is hoped by the authors that this might inspire some deeper academic research into the systemic aspects of farming. The insights developed by the authors have been driven from the very beginning by the analysis of empirical evidence. The MSO model emerged in two parts: an explanation that fitted the evidence and the subsequent search to establish why costs were behaving in an unexpected fashion.

August 2025

ABOUT THE AUTHORS

Chris Clark is a non-generational farmer, who when at agricultural college with his wife-to-be, decided that they would buy a farm before he turned 50. His post-college career, with his wife, included spells farming entirely free-range pigs on an organic arable farm, as a tenant farmer and setting-up a successful management consultancy practice specialising in marketing and graphic design work. He purchased Nethergill, a 170 ha farm set in the magnificently moody landscape of Upper Wharfedale, Yorkshire, at 49 and set about turning it into a model of upland farming applying all his considerable management consulting experience.

Brian Scanlon is a graduate in mathematics and physics with a post-graduate qualification in operational research. His career started in the steel industry and later embraced management consulting and banking. He had, in parallel, set-up a company which had effected some privatisation projects and flotations on AIM. Farming, to him, was an alien world and he would be 55 before he set foot on a farm. His career had been with major corporations in North America, Europe, and Japan.

Nethergill Associates

Nethergill Associates is a North Devon, farm-based business management consultancy. Throughout the book frequent reference will be made to the Nethergill Associates database. This comprises up to 150 datapoints for each farm that has been critically examined. By mid-2024 this database incorporated 320 farm business accounts. From time-to-time sub-datasets will be referred to where conclusions need to be drawn for, say, a district, a landscape or a sector such as dairy farming. The whole approach of the authors in developing the MSO theory has been based on the empirical evidence provided by the database.

www.nethergillassociates.co.uk

ACKNOWLEDGEMENTS

The Profitable Farm would have been so much harder to complete without the practical help and advice we received. So, many, many thanks to: Fi Clark for her unequivocal support and down-to-earth impeccable judgement; Clare Fynn, behind the scenes, providing invaluable planning and organisation; Dr Andrew Hattan, farmer, cheesemaker and expert ruminant nutritionist who was from the start of our journey a practical and scientific sounding board; and Professor Tim Benton, another scientific sounding board, keeping us on the straight and narrow.

Developing the MSO theory would not have been possible without the wonderful farmers we at Nethergill have had the pleasure to work with in recent years. Their financial and yield figures populated the database and fed the algorithm behind MSO. Their openness to a different way of thinking encouraged us no end.

Finally, thanks to the Nature Friendly Farming Network, the National Trust, the RSPB and the Wildlife Trusts for the funding and support behind the 'Less Is More' and 'Farming at the Sweet Spot' reports out of which this book grew.

ABBREVIATIONS

AT	assets turn
BNG	biodiversity net gain
BPS	Basic Payment Scheme
CA	current assets
CAP	Common Agricultural Policy
CL	current liabilities
CVC	corrective variable cost
ELMS	Environmental Land Management Scheme
FA	fixed assets
FMCG	fast-moving consumer goods
LSU	livestock unit
LTL	long-term loans
MSO	maximum sustainable output
NPV	net present value
OF	owner's funds
P&L	profit and loss
PAT	profit after tax (and interest payments)
PAYE	pay-as-you-earn
PBIT	profit before interest payments and tax
PE	price to earnings ratio
PI	primary income
PNC	primary natural capital
PVC	productive variable cost
RE	retained earnings
ROTA	return on total assets
S&A	source and appropriation of funds statement
SBU	strategic business unit
USP	unique sales proposition

Chapter 1
INTRODUCTION

1.1 GENESIS

Nethergill is a 170 ha farm set in the magnificently moody landscape of Upper Wharfedale, Yorkshire in England. It was purchased by Chris Clark, one of the authors of this book, in 2005. Chris set about turning it into a model of upland farming applying all his considerable management consulting experience.

All his efforts failed to create a profitable farm business and it survived on bed & breakfast bookings, a holiday lets business and the support and subsidies available to farmers. He then started to downsize the farm business and found that the farm started to become more profitable. It defied all logic and the conventional wisdom of standard economic theory. He invited an old friend, Brian Scanlon, the other author of this book, to visit the farm in 2017 and have a look at its business model.

1.2 A PARADOX

Other farms in the district, which had downsized more modestly to adjust to the prevailing economic conditions, also reported an improvement in profitability. These reports seemed to fly in the face of business logic and were often dismissed. Many of the farmers expressing these views, including those who were practising regenerative farming, were by their own admission unable to quantify any changes with any confidence and were uncer-

Figure 1.1 Nethergill Farm.

tain as to how true profitability should be measured anyway. They all just felt they were better off.

Eventually, there were too many similar stories to ignore, and some were being voiced, quietly, by farmers that were measuring their business performance with some confidence. If the experiences were valid, it raised the question as to what sort of mechanism was in play and what was driving it. The phenomenon could not be explained by any of the business models adopted by the farming community, and this was a great concern.

Initially, Brian thought that it was simply not possible to become more profitable by downsizing unless it was entirely driven by reductions in fixed costs. However, this had not been the case at Nethergill. There had to be some pattern of behaviour, not recognised by the standard economic models, that was responsible. This raised a number of questions, such as, why should food production be different from other fields of economic endeavour?

At a session around the large farmhouse table looking out over a wet cobbled yard the metaphorical head-scratching ended with the recognition that food was a fuel and fuels were a form of energy. Perhaps by looking at farming entirely from an energy standpoint it was thought that some new insights might emerge.

To explain, empirically, what seemed to be happening a variable cost line that comprised two sequential components (technically, a bifurcated line) did the job, but why would such behaviour be warranted? The insight came with the sudden realisation that farming had the challenge of orchestrating two very different types of energy. There was the energy of direct sunlight and the industrial energy embedded in additives and substitutes, such as fertilisers and feed concentrates.

Farming practices have changed significantly from the time of the First World War. The two great wars of the twentieth century focused attention on maximising food production from domestic resources, essentially at any cost. Then, it was a matter of national survival as shipping lanes came under attack. The driving force was to maximise the land available for food production and make continuous improvements to yields to maximise outputs. Science and industry was mobilised to help particularly in terms of agricultural-chemicals, factory systems, mechanisation and intensification practices. It was a great success.

During this time, little thought was given to aspects of profitability or the environment, and funds were made available to maintain output. In the years following the Second World War, this pattern was encapsulated in two beliefs that were taken to be axiomatic: that support payments were unavoidable if farming was to survive and that no farmer could go wrong in producing more food.[1]

These sentiments, probably, still prevail today for many farmers.

1.3 THE SIGNIFICANCE

If farm profits are not maximised when farm outputs are maximised then some subtle trade-off is being missed in the calculations being applied. Turning it around, if some farms do become more profitable by downsizing it must also mean that these farms become more unprofitable as outputs are pushed forward.

All unprofitable businesses place a burden on the economy that will become unsustainable at some point. In farming, this is often masked by the practice of support payments and the informal subsidies that come from other sources of income in the family, which are used to maintain a much-valued lifestyle.

In parallel with the economic aspects of sustainability there is also the question of environmental sustainability. Farming, it is now widely accepted, has caused and continues to cause significant damage to the environment. This raises the spectre that not only might farming be economically unsustainable, but it is also unsustainable environmentally in its present form. Could there be some limit to farming practices that would represent the **maximum sustainable output** (MSO) on a farm in a way that addresses both economic and environmental issues?

To examine this, we will make constant reference to seven themes that run through this book; the biosphere; the managed landscape; the unavoidable physics of energy; the concept of natural capital; the importance of profitability in businesses; the randomness of Nature; and the issue of coexistence in a shared ecosystem.

1.4 THEMES

The biosphere

The biosphere is the place where life interacts with the atmosphere, the soil and bodies of water. It comprises organisms and plants that interact through mechanisms of photosynthesis, respiration and decay.

Animals extract oxygen (O_2) from the air and exhale carbon dioxide (CO_2) through respiration. Happily, plants extract oxygen and carbon from the carbon dioxide in the atmosphere, hydrogen (H_2) from water, and nitrogen (N_2) from soils, and then exhale surplus oxygen.

Dead animals and decayed plants are re-cycled through the soil. These inter-connexions between animal life and plant life are essentially encapsulated in the C-cycle (carbon), the N-cycle (nitrogen) and the water cycle.

Carbon, and its distinctive form of organic chemistry, is central to life on Earth. In the C-cycle, plants absorb sunlight and, through a process of photo-synthesis, convert the carbon dioxide in the atmosphere into sugars, fats and proteins (Figure 1.2). Photosynthesis is the key, and it is of such fundamental importance that the chemical equation, known as the Calvin-cycle, is worth setting out.

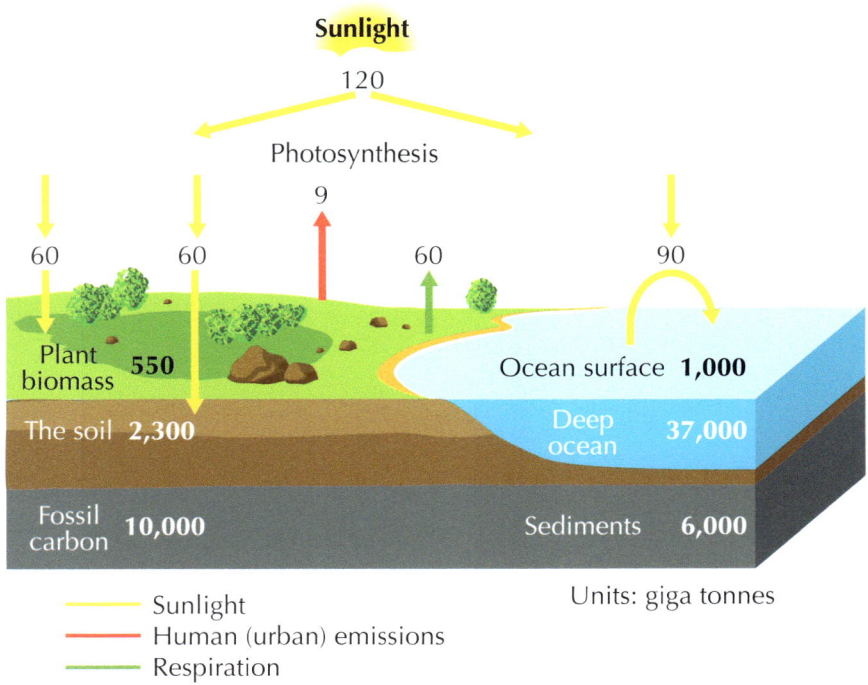

Figure 1.2 The carbon cycle.
Source image: AdobeStock_45383066. Illustration: Elaine Leggett. Data: US DOE/NASA

$$CO_2 + 2H_2O + \text{light energy [photons]} > CH_2O + O_2 + H_2O$$
carbon dioxide + water + sunlight > sugars + free oxygen + water

Other organic matter, in its abundant forms, will lock-in the atmospheric carbon compounds in different ways. It takes hundreds of years in trees whereas it takes tens of millions of years in hydrocarbon strata. All this can, of course, be un-locked instantly on combustion of the fuel compounds of carbon, such as methane (CH_4).

$$CH_4 + 2O_2 > CO_2 + 2H_2O$$
methane + oxygen > carbon dioxide + water

Nitrogen is the most abundant element in the atmosphere, accounting for 71% of its volume. However, it is notoriously difficult to synthesise compounds of nitrogen. In the N-cycle, plants extract from soils the different forms of nitrates (compounds of nitrogen with some other element and an abundance of oxygen) produced by bacterial action (Figure 1.3). Decay produces ammonia (NH_3) and other bacterial action will first produce nitrites (compounds of nitrogen and some other element with lesser amounts of oxygen) and then the nitrates needed for plant life and animal foods. This bacterial action produces about 95% of the requirement for plants. The remaining 5% is the consequence of lightning in the atmosphere and the rainfall that washes this into the soil.

The role and the impact of the water cycle is very different (Figure 1.4). Water is essential for all plant growth. The essential source of water for the soil is rainwater. This is simply the precipitation that results from the evaporation of water from lakes, seas and oceans by convection currents. A secondary source is provided by irrigation schemes that channel water to places where it is needed. Where water supplies are not replenished naturally there is a growing tendency to extractively pump groundwater from aquifers. This source will be limited, and when it is exhausted, it might not be replaced for millennia.

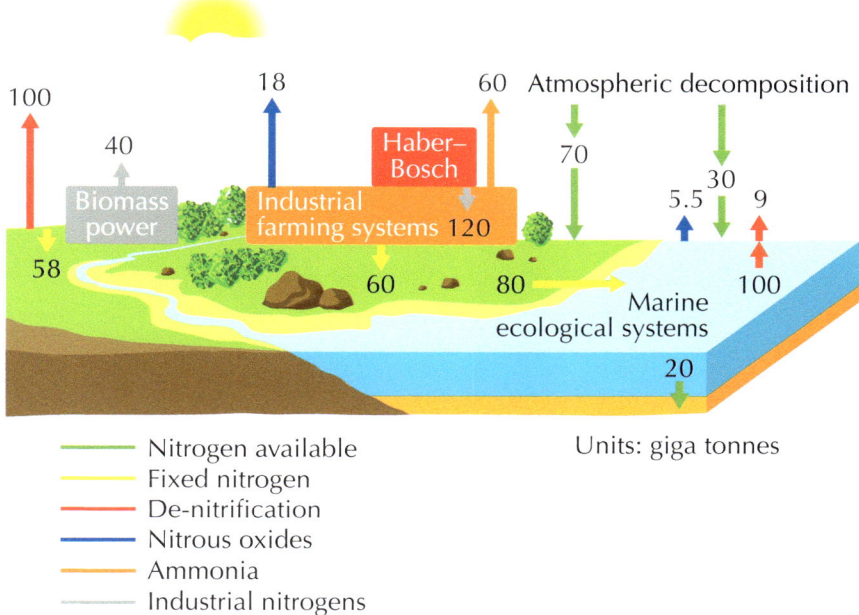

Figure 1.3 The reactive nitrogen cycle.
Source image: AdobeStock_45383066. Illustration: Elaine Leggett. Data: US DOE/NASA.

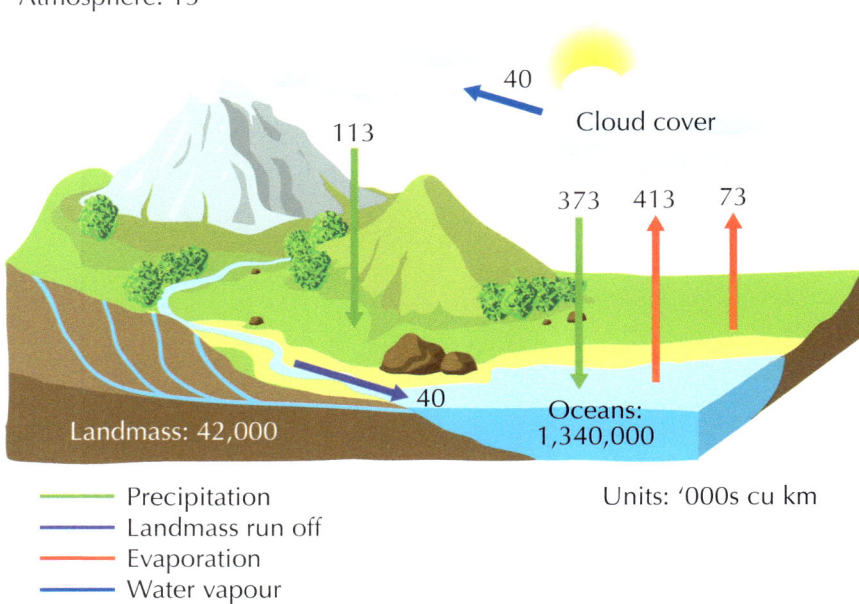

Figure 1.4 The water cycle.
Source image: AdobeStock_45383066. Illustration: Elaine Leggett. Data: Royal Meterological Soceity.

The managed landscape

The managed landscape we enjoy is a consequence of the realities of geography, such as latitude, elevation, precipitation and specific phenomena, such as the Gulf Stream, etc. It will differ from place to place. These accidents of geography are the driving forces behind attempts by farmers over the millennia to take corrective action to compensate for any perceived supply or commercial disadvantages. This is the origin of the managed landscape. In the UK, as in many other countries, the managed landscape is now the *de facto* natural landscape in the absence of any prime wilderness (Figure 1.5).

The Saxons, between 500 and 850, were responsible for much of the original heavy ploughing and passed on the practice of rotation with their three-field system. The aristocratic Normans, from 1100, focused more on woodland management by clearing areas (the process of *assarting*) for cultivation and retaining forests for hunting. The monasteries, from 1250, pursued upland colonisation and created the foundations of the mediaeval wool trade while some of the great cathedrals, such as Glastonbury, Canterbury, Peterborough and Ely, promoted, from 1400, programmes of marshland and fenland

Figure 1.5 The managed landscape.
Credit: Alamy 2HBN003.

recovery. The great estates in post-Cromwellian times, particularly from 1720, continued with woodland clearances and, through access to power in Parliament, introduced a practice of enclosures. This was ahead of its time in a European context and greatly improved agricultural productivity, but not without severe disruptions in many agricultural communities and the displacement of many people. In England, the flight from rural areas instigated by industrialisation, from 1750, helped to absorb the displaced rural population. In the Scottish Highlands and in Ireland the absence of industrialisation prompted mass emigration to the New World.

This Age of Enlightenment attracted the attention of inquiring minds to agricultural improvements and the science of the day was applied to crop rotation (by Townshend), selective breeding (by Coke), and mechanisation (by Tull). The islands of the Caribbean were exploited at this time, too, to compensate for what was termed 'the lost acres' in a reference to the superior agricultural resources of France, the formidable international competitor to Great Britain at the time.

The intervention of the UK's early and vigourous industrialisation at this point had a dramatic influence. Food supplies came under greater pressure, which placed more attention on overseas sources, and mainstream science became more interested in industry than in farming. Up to this point, the landscape had been 'nudged' along with small incremental changes. When, some 150 years later, extensive guano deposits were discovered in Chile it promoted a greater interest in the industrialisation of farming by forcing improvements in yields. Germany, which was also rapidly industrialising at this time, looked to its new chemical industries to produce artificial fertilisers and, with the development of the Haber–Bosch process to produce ammonia, it changed the dynamics of farming in industrial societies. The advent of the First World War, which interrupted access to guano, heralded-in a new age of intensive farming.

The era of intensive farming introduced changes to the landscape on a much faster timetable than previously and this has led to many of the environmental challenges we face today. After the First World War agriculture attracted more and more state involvement. Marketing boards were created in the inter-war years to manage food prices and supplies and our later accession to the European Community exposed the sector to its Common Agricultural

Policy (CAP). Today, farming not only resides almost entirely within a managed landscape but the sector is heavily dependent on the state and its political objectives (Figure 1.6).

The UK's withdrawal from the European Union and its CAP regime will change farming in the immediate future. The Basic Payment Scheme (BPS) of the CAP is currently (2024) being phased out and being replaced by the Environmental Land Management Scheme (ELMS) on a progressive basis. The essential difference allows the possibility of payments related to public goods in contrast to payments on an area basis. In particular, food security can only be guaranteed when farm businesses return to being intrinsically profitable enterprises.

The unavoidable physics of energy

Food is a fuel and fuels are a form of energy. The physics of energy are unavoidable and will apply to farming just as much as it does to any other energy producer. Energy can be neither created nor destroyed; it simply gets converted from one form to another. However, every time a conversion process takes place energy is lost and this lost energy will be irrecoverable.

> To illustrate this point, consider two ways of using gas from the North Sea for cooking in the kitchen. One way is to pipe the gas directly to the home and have a gas-fired cooker. The other way is to pipe the gas to a (gas-turbine) power station, transmit the electricity generated via the National Grid to the home, and use it to run an electric-element cooker. The second method has at least two more steps in the conversion chain and is significantly more wasteful as a result.

Now compare this with the use of energy on a livestock farm. One way is to use sunlight to grow forage, which is then eaten by ruminants to produce meat for human consumption. Another way is to buy-in artificial food concentrates (full of calories) to feed the livestock to produce meat. The food concentrates in this second method will require manufacturing and transportation and consume industrially produced energy in the process. Also,

Figure 1.6 Timelines influencing agriculture.
Credit: Nethergill Associates. Illustration: Elaine Leggett.

the real value of the ruminant, with its four stomachs, is being wasted by using it only as a mono-gastric beast. Again, the additional conversions (and the failure to benefit from freely available sunlight) make the second option much less attractive. As will be seen later, this option is not only going to be more profligate in energy terms, but it cannot deliver a true net benefit.

The concept of natural capital

The landscape is a capital resource and will deliver benefits for society in the form of Nature's bounty. Farm properties produce goods for consumption by mobilising this capital in the same way that a factory is a capital asset that can produce manufactured items. A manufacturing business will typically be owned by its shareholders, that is those people who have subscribed monies to purchase productive assets, to provide the funds to purchase supplies and to pay for productive labour ahead of income from sales. Although a farm property may have been purchased by a farmer it is Nature that provides the critical productive asset. In the way that industry acknowledges a liability to its shareholders (by justifying the use of its shareholders' funds) so a farmer might acknowledge a corresponding liability to Nature and treat it as a stakeholder in a farm business.

This raises fundamental questions about natural capital and how it might be valued.

The importance of profitability in businesses

Farming is such a technical activity with demanding challenges presented by the weather, animal health issues and crop diseases that the purely business aspects can be easily neglected. Compared with the outdoor farming life, running a business can fade into the background as a priority. However, a farm is first and foremost a business.

All businesses will reflect an underlying **mission**. In farming, the mission is to use the natural resources that come with owning and working the land to produce food and other valuable natural products. In doing this the farm is satisfying some of the needs of the **market** for food and other goods. This produce will command a price in the marketplace, which is set by mechanisms that match supply and demand for goods of a particular type and quality.

All businesses need an **objective**. In general, the objective of a business is to make a **profit**. This profit should be sufficient to maintain the standard of living and the desired lifestyle of the farm owners. Unprofitable businesses will simply disappear when the cash resources of the owners run out. Profitability is an essential prerequisite of viability.

All businesses have to acknowledge some wider responsibilities within the community for the sake of good order. Driving profits cynically without due regard for others has no place in good business management. In farming, there are two paramount responsibilities to discharge. There is an implied duty of care to animals and **animal welfare** must be uncompromised. In working the landscape, farmers also have a responsibility to avoid degrading the **environment**. Environmental damage not only affects a community, but it also destroys the business value of the very natural resources on which farming depends.

Today, farming resides in a **managed landscape**. The natural prime wilderness disappeared long ago. This managed landscape is in a state of dynamic equilibrium with Nature. Changes have been made, in small increments, over the centuries and Nature has adjusted to accommodate these. This has happened to an extent that it is quite valid to argue that the managed landscape is now the natural landscape.

In recent decades farming has undergone changes at an unprecedented rate when considered from a long-term perspective. The availability of artificial products, such as fertilisers, herbicides and pesticides, have been mobilised in the cause of securing better yields; mechanisation has reduced the burdens of manual labour and transformed farm outputs; and science has been applied to issues such as selective breeding and genetic modification. The speed of this recent change has put new stresses on the landscape as Nature tries to adjust.

Additionally, in both national and community terms, farming has a **role** to play and its role comes with two obligations. It is a custodian of the environment with a duty of care and it is a beneficiary of Nature's bounty with duty to turn it into an economic benefit for the community. A failure to meet this second obligation would reduce farming to a role of park-keeping or simply gardening.

The randomness of nature

Consider what happens when a managed landscape, which has been maintained in a state of dynamic equilibrium with Nature, is simply left unattended. Nature will appropriate the natural sources of energy formerly employed in farming. It will evolve randomly down a path of least resistance towards a new point of equilibrium. Until this point is established, which could take decades or even centuries, two phenomena emerge.

A dominant species of plant or fauna will emerge to exploit its new-found advantage and will prevail until other species respond in competition.

From time to time, a mixture of aperiodic influences (repetitive-type behaviour in an irregular time-cycle) and positive feedback (exponential growth leading to eventual species collapse) introduce a chaotic effect. However, chaos can produce new forms of order through a phenomenon known as **resonance** and this tends to become a singular feature of the new landscape. The result is surprise and uncertainty (the, so-called, *law of unintended consequences*).

These are the reasons why so many re-wilding initiatives fail or disappoint their supporters. Desirable patterns of biodiversity rarely emerge from habitats specifically designed to promote those very patterns.

The issue of coexistence in a shared ecosystem

The biosphere is the world we share; the managed landscape is the history we share; the unavoidable physics are the chains and constraints we have to endure; and natural capital is the bounty we enjoy. Together, these worlds combine in an ecosystem, which is the life we share.

Life is a pyramid of coexistence (Figure 1.7).

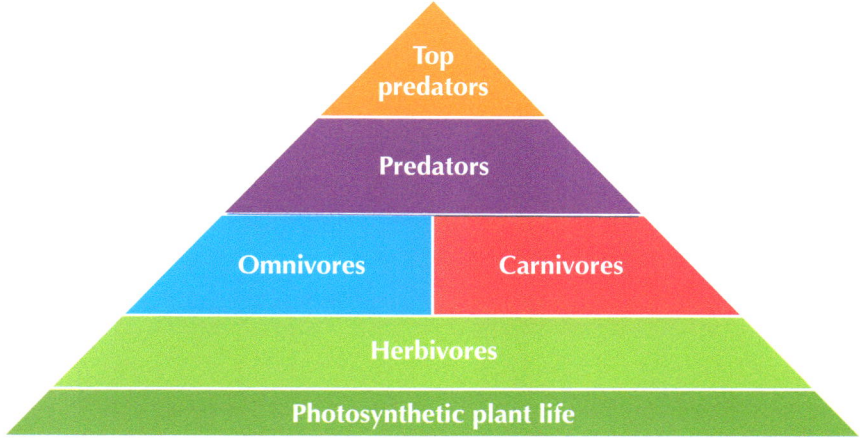

Figure 1.7 Hierarchy of predatory species.
Credit: Nethergill Associates. Illustration: Elaine Leggett.

Top predators sit over lesser predators; these sit over the carnivores (and omnivores), which in turn sit over the herbivores. At the bottom of this pyramid are the producers – the photosynthetic plant life.

Maintaining this pyramid in a credible, balanced and un-disturbed form is not only a key responsibility of farming – it is an essential prerequisite of human survival. It requires uncompromised standards of animal welfare, the maintenance of perpetual fertility in the soil, a proper balance between commercial and conservation imperatives, and the accommodation of continuous changes to our climate.

1.5 THIS BOOK

This book is effectively the story of how the concept of maximum sustainable output (MSO) evolved and how it prompted excursions into many adjacent fields of endeavour. Its purpose is twofold: first, to help farmers become more profitable without compromising good farming practice; and second, to influence policymakers in their roles to incentivise farmers to pursue greater profitability. It was inspired by some of the issues raised when the authors set out to try to explain, or disprove, the phenomenon of some farms becoming more profitable when downsizing and reducing yield. This quest was to lead to some surprising and counter-intuitive conclusions – more, of which, later.

This book is not an academic textbook nor is it a prescription for better farming. It details a journey in which problems were, initially, examined empirically to establish causal relationships and then analysed, mathematically in retrospect, to introduce predictive models. **Its underlying thesis is that energy considerations will drive all outcomes**.

The seven themes set out above are ever-present in this book. However, they will not run conveniently in a logical sequence but will develop more along parallel lines. Many of the perspectives in this book are new in a farming context and some of the concepts discussed have required some new terminology. None of the mathematics, physics or economics in this book will be beyond a competent farmer well-suited to being in business.

Some of the terms, such as those relating to MSO, productive variable costs (PVCs), and corrective variable costs (CVCs), are entirely new and have been coined directly as a consequence of the authors' work. A comprehensive glossary can be found at the end of the book and should be referred to frequently until the different terms become familiar.

The book is organised into informal sections to give readers a choice in addressing its contents.

Chapters 2 to 5 tell a story. Chapters 6 and 7 are technical and some readers may want to skip these and return to them later when they will have a better overview. Chapter 8 comprises case studies and should act as a recap when the book is consulted from time to time after being properly read. Chapters 9 and 10 attempt to put farming in a wider political, economic and social context.

Some re-reading of earlier chapters will be inevitable.

Chapter 2
FARM ACCOUNTS

2.1 STATUTORY ACCOUNTS

All businesses are required to file annual returns with the authorities. These returns are the statutory accounts of the business and should comprise three key documents: a balance sheet (often referred to as the B/S); a profit and loss account (often referred to as the P&L); and source and appropriation of funds statement (often referred to as the S&A). In practice, few of the statutory accounts that are filed by farm businesses will include an S&A.

The purpose of the returns is twofold. The first is to ensure that the funds subscribed by investors have been properly used and accounted for. This is very relevant in the case of publicly quoted companies. The second purpose is to ensure that the taxation liabilities of the business have been properly assessed.

The tax authorities go to some length to ensure that only *fair* profits are taxed and, in doing so, cater for various allowances and concessions to be made. Certified accountants, in preparing the returns, are effectively guaranteeing the honesty of the returns to the tax authorities. However, the often-complex adjustments to figures from the underlying business can result in these statutory accounts being a poor tool for management control purposes. Far too many farm businesses just ignore these accounts, confessing to the B/S being a complete mystery and to being invariably unaware of the S&A.

2.2 THE NATURE OF MONEY

Money would seem to be a simple thing to understand. We have notes and coins to work with and these have a clear **value**. However, money has another aspect – **term**; and this is often forgotten. Term reflects the time-aspect of money. Money in your pocket for a week before spending will be different from the same amount on deposit in a building society for a number of years. This makes it surprisingly complicated.

Consider the problem of sailing across a river from one side to the other. Where you end up will be determined by the direction and force of the wind and the flow of the river or the tide. Your actual passage will be some combination of the two effects. Accounting for the two aspects of money (value and term) was recognised as a problem by the great banking families, such as the Medicis, in the small Italian city-states in the fifteenth century. Their solution was to adopt two complementary sets of accounts, and these are encapsulated in the B/S (to address the term issue) and the P&L (to address the value issue). The B/S for one year-end is related to the next year-end B/S by the P&L. The S&A is a summary of the money flows between the B/S and the P&L and is analogous to a cash-flow statement. It was an inspired piece of work, and we still benefit from these conventions today.

2.3 THE BALANCE SHEET (B/S)

A balance sheet is a statement that reconciles two features of a business:

- the physical and financial resources that comprise the business such as land, buildings, plant, machinery, and cash;
- and the funds made available to the business to purchase the necessary property and productive resources.

The physical properties are defined as **assets** and the funds are defined as **liabilities**. The assets and liabilities in all businesses must be accounted for in full and should, therefore, balance each other (in terms of sterling or other currencies).

Historically, a B/S comprised two parallel columns: assets in the left-hand column and liabilities in the right-hand column. Today, it is more usual to present a B/S in one continuous column with the assets being placed above the liabilities.

Assets

Assets form two groups on a B/S. These are the fixed assets (FA) and the current assets (CA).

FA represent the long-term physical resources of the business, which may not be turned back into cash very quickly or easily, and will typically comprise:

- land
- buildings
- plant and machinery
- investments (in other companies or securities).

CA represent the short-term resources of the business, which are simpler to turn back into cash and will typically comprise:

- cash
- accounts receivable (that is monies owed by other traders: the *debtors*)
- stocks and inventories.

When profits are made in a business they will be used to reward those who have put funds into the business. As the funders (investors) are accounted for as liabilities, all retained profits are treated as liabilities. Many find this counter-intuitive because our language suggests that assets are good and liabilities are bad. On a B/S, assets and liabilities are neither good nor bad; the convention simply reflects the premise that assets have to be acquired responsibly in deference to the risks taken by the funders.

Liabilities

Liabilities form three groups on a B/S (Figure 2.1). These are the owners' funds (OF), any long-term loans (LTL) and the current liabilities (CL). The owners' funds represent the original capital sums put up by the investors to get the business started together with any retained profits from the growth of the business.

LTLs comprise monies borrowed on a term in excess of 1 year and will include bank loans and mortgages. Businesses have a liability to discharge these loans in the proper time.

CL represent the short-term (less than 1 year) calls on the monies in the business and will typically comprise:

- bank overdrafts
- accounts payable (that is monies owed to other traders: the *creditors*)
- miscellaneous liabilities, such as interest payments, taxes and dividends due.

Goodwill

In time, it gets more difficult to make both sides of the B/S balance in actuality. Accidents happen, losses can be made and assets change in real value. When the aggregate liabilities exceed the assets value (in the happy circumstances that good profits are made and retained or the unhappy circumstances that poor quality assets have been acquired), accountants will define the difference as 'goodwill' and allow the B/S to balance. Goodwill is regarded as an intangible asset and is used to account for the implied value of important but non-physical aspects such as contracts, licences, patents and reputations, etc.

Interpretation

The B/S is a justification of the way in which monies have been spent wisely and represents a snapshot at a particular point in time. A B/S is traditionally

Figure 2.1 Simplified B/S for a typical farm business

		£	£
ASSETS			
Fixed assets	Land	250,000	
	Buildings	85,000	
	Plant	30,000	
			365,000
Current assets	Cash	15,000	
	Accounts receivable	40,000	
	Livestock	210,000	
	Stores	20,000	
			285,000
Intangible assets			200,000
TOTAL ASSETS			**850,000**
LIABILITIES			
Owners' funds	Capital investment	40,000	
	Retained profits	75,000	
			115,000
Long-term loans	15 year mortgage	600,000	
	5 year bank loan	50,000	
			650,000
Current liabilities	Bank overdraft	50,000	
	Interest payments	5,000	
	Accounts payable	30,000	
			85,000
TOTAL LIABILITIES			**850,000**

compiled every year on a specified date (the reporting date for the business). Comparing a B/S with its successor measures the term aspects of the business. The mechanism to get from one B/S to the next is encapsulated in the P&L.

2.4 THE PROFIT AND LOSS ACCOUNT (P&L)

The P&L will be specific to the period between two successive B/Ss. It is a record of the revenues earned, the costs incurred and the consequential surpluses or losses for this period. Most statutory P&Ls will carry the minimal amount of data consistent with meeting the requirements for filing an annual return (Figure 2.2). This is to protect information that is personal or commercially sensitive, but it restricts the utility value of these accounts in controlling business activities.

Revenues

The revenues in a business will mostly come from the sale of goods and produce. This will be the value of the farm output from all sources.

For some time now, the farming sector of the economy has drawn significant levels of subsidy from the government and other agencies in the form of

Figure 2.2 A simplified P&L for a typical farm business.

	£	£
INCOME (REVENUES)		
Sale of goods and produce	220,000	
Support payments	40,000	
Grants	15,000	
		275,000
EXPENSES		
Cost of sales	130,000	
Administration	100,000	
		230,000
PROFITS		
PBIT		45,000
Interest payments	20,000	
Tax	4,500	
PAT		20,500
Retained earnings		10,000
TOTAL ASSETS		**850,000**

support payments and grants. Strictly speaking, these are legitimate revenues and are part of the calculations of profit and consequent tax liability. However, as will be examined later, these sources of revenue have no bearing on the underlying operational profitability of a farm.

Cost of sales

The direct costs incurred in producing goods for sale, such as purchases of supplies, casual labour and veterinary expenses, are regarded as the cost of sales, and treated, usually, as a **variable cost**. That is, these costs are taken to be directly related to the volume of output in the sense that if output volumes increase these costs will increase in proportion and if output volumes fall then the reverse will prevail.

Administrative costs

These are the costs that are related to running a business establishment and will be incurred whatever the output volumes might be. As such, these costs are regarded as a **fixed cost** in the business. These cost items will include rents, rates, utility charges and professional fees.

Drawings

Farmers will draw monies out of their businesses for both their work done and as reward for their enterprise. For incorporated businesses, drawings (usually in the form of salaries) are recognised as a business cost in the calculation of profits. In these cases, the tax liabilities on salaries are a personal matter and will be subjected to the pay-as-you-earn (PAYE) conventions. Any surpluses will then be subjected to corporation tax.

For true partnerships and sole traders, the drawings will comprise all the distributable income after expenses and adjustments. As the share of this income for each partner will then be treated as a personal tax liability, farmers have to manage their cash flows to ensure that tax liabilities can be met at the appropriate time.

Profits

Profits are simply a measure of the differences between revenues and costs. However, profits in a set of statutory accounts are expressed in several different ways in the P&L:

- PBIT: Profit before interest payments and tax. This is actually the taxable profit for the year in question. Bizarrely, because of adjustments in the form of accruals, provisions, deferments and prepayments a profit may be reported when the cash surpluses in the business may be zero. Confusingly, in the statutory accounts of many farm businesses interest payments are treated as part of the fixed costs of the business.
- PAT: Profit after tax and interest payments.
- RE: Retained earnings, which are the amounts transferred to the B/S to become part of the owners' funds (OF).

Cash flow

If losses are incurred, monies will have to be transferred from the B/S to cover the deficits. To do this, assets must be sold or cashed-in and if this continues indefinitely, eventually all the assets in the business will be liquidated. Alongside the P&L it is usual to prepare statements of cash flows in the form of an S&A. These are not strictly enforced as statutory requirements, but they are critical in tracking movements between the B/S and the P&L when capital investments are made or when operational losses are covered.

The S&A will relate to the same 1 year period as the P&L. It should not be confused with an exercise to produce *cash-flow projections*. These are vital in the short-term management of a business and should be done frequently to ensure that a business maintains its liquidity.

The simplified set of accounts in Figures 2.1 and 2.2 do not reflect the considerable level of adjustments that are introduced by accountants. As certain assets deteriorate with age and use, such as tractors and similar plant, the tax authorities allow depreciation charges to be set aside, free of tax, to replace these worn assets with the idea that the depreciation account will be sufficient when called on (but this is rarely the case, in practice). There are usually considerable adjustments to ensure that revenues and costs are assigned to the appropriate period through the conventions of accruals and deferred payments. These adjustments can heavily distort the differences between reported profits and actual operating profits. This makes the statutory accounts confusing for the purposes of management control and this confusion is exacerbated when farm businesses adopt accounting packages that produce these accounts on a monthly basis.

All businesses should seek to minimise their tax burden by legitimate means. Tax avoidance is legitimate but tax evasion is illegal. Good accountants will offer advice on mitigating tax liabilities. Attention is often focused on buying plant or equipment as a way of reducing taxation, but this practice can backfire badly. Equipment, to be cost effective, needs to be utilised to a high degree. However, the seasonal nature of farming will usually preclude this situation and too often farms end up with under-used, over-specified equipment that attract significant repair and maintenance costs. Profit is a worthy objective and sometimes it is simply better to pay taxes rather than avoid them.

2.5 MANAGEMENT ACCOUNTS

In the P&L shown in Figure 2.2 it should be noted that while the expenses in the business amounted to £230,000 the revenues from sales (farm output) were only £220,000. Without grants and support, this farm would have returned an operating loss of £10,000. This indicates the need for a separate set of accounts that are constructed in a way that helps the effective management and control of the business. These are the management accounts.

Management accounts are ordered in a very different way so that key evaluations can be made. In conjunction with this, the concept of contributions is introduced to substitute for profits and the different contributions, measured at different points (or levels), are taken to be the true measures of business profitability (Figure 2.3).

In management accounts, income is taken to be the revenues generated from sales of goods and produce (farm output). All support payments and

Figure 2.3 Management accounts for a typical farm business.

		£	£
FARM OUTPUT			
Sales of livestock	Cattle	140,000	
	Sheep	60,000	
	Winter wheat	20,000	
			220,000
VARIABLE COSTS	Casual labour	15,500	
	Concentrates	38,500	
	Fertilisers	54,500	
	Veterinary	12,000	
	Bedding/sprays	9,500	
			130,000
1st contribution		**90,000**	
FIXED COSTS	Rent and rates	10,500	
	Full-time labour	50,000	
	Utilities	4,000	
	Fuel and oil	5,500	
	Office expenses	12,000	
	Legal fees	10,000	
	Accounting fees	8,000	
			100,000
2nd contribution		**−10,000**	
SUPPORT PAYMENTS	BPS	30,000	
	Environmental	10,000	
			40,000
3rd contribution		30,000	
TOTAL ASSETS EMPLOYED		850,000	
		Margin (%)	Assets turn
ROTA	Pre-support	−4.55	0.26
	Post-support	15.38	0.31

grants are regarded as business bonuses and accounted for after all operational costs have been deducted.

In farming businesses there are three fundamental levels of contribution to consider.

1st contribution

This is defined as:

1st contribution = Farm output less variable costs

This is the financial contribution available to cover all subsequent costs and calls on income after the variable costs of production. If this is negative, the business will lose cash on every transaction and will be intrinsically unviable. However, farms in a state of transition will often exhibit this behaviour but these farms should have the prospect of full profitability in due course.

2nd contribution

This is defined as:

2nd contribution = 1st contribution less fixed costs

This is the contribution available for drawings, profits, capital expenditures, interest payments and taxation. If this is negative, following a positive 1st contribution, a business will be decapitalising as monies are drawn from the balance sheet. Fixed costs can be a considerable burden for small farms and are a major source of unprofitability.

3rd contribution

This is specific to farming, and is defined as:

3rd contribution = 2nd contribution plus support payments and grants

This will approximate to PBIT in the statutory accounts. Although the majority of farms would appear to be profitable on this basis their ability to deliver adequate drawings, which are taken out at this level, is typically poor. Few small farms support a decent standard of living for a family and a second income is often an imperative.

Support payments are entirely political. Successive governments and the policies of the CAP of the EU, in the past, have reflected a fear of the impact of real food prices on the electorate. This support regime is already being replaced progressively in England by ELMS. In some academic circles it is expected that farming will move away entirely from support mechanisms in the future. To accommodate this, farm prices would inevitably rise but consumer resistance can be expected. This will put farm profitability issues under much greater scrutiny.

The economic sustainability of farming depends not only on making a profit, but as will be seen later, it depends also on making a profit that is adequate to support a family and deliver a return on the capital funds at risk in the business.

The particular detail that will comprise the management accounts will differ from business to business. In time, the most useful revenue and costs categories emerge as farmers improve their ability to frame these accounts and exercise greater control. As management accounts are not part of the statutory accounts, and can always be kept confidential, farmers should embrace as much detail as possible in the interests of a better understanding and control of activities. There is a powerful maxim in industry that encapsulates the utility of management accounts: *If you can't measure it, you can't control it.*

2.6 CASH-FLOW MANAGEMENT

The management of cash flows in a business is probably the most important of all business management disciplines. In all farm businesses, this should comprise monthly projections for at least 6 months ahead on a rolling basis. Ideally, the control documentation should comprise four distinct sections.

Cash-flow projections for trading activities

In simple terms this comprises tracking:

- income from sales
- operational expenses
- interest payments

- hire-purchase/ leasing expenses
- mortgage payments
- projected tax liabilities
- operating balances (= income less expenses).

Cash-flow management is essentially about discipline and anticipation. Have the discipline to set aside monies for the calls that will come. There is nothing more effective than a 'cookie-jar' accounting model. Put, notionally, a series of jars on the mantlepiece marked 'Tax', 'PAYE', 'Rents', 'Mortgage Payments', etc. and set aside monies regularly for each call. The balances left for living expenses may well fall but there will be a better peace-of-mind as fewer nasty surprises emerge.

If cash flows are negative they must be covered by bank deposits. A long run of monthly deficits projections must be addressed quickly. Unless there are real prospects for a future stream of income this will be the first signal of an unviable business. If a future income stream is expected, short-run deficits may be covered by an overdraft facility. As a management discipline, overdraft balances should be cleared at the earliest opportunity. Working permanently to overdraft limits is bad practice: it is expensive, and it denies the business any room for manoeuvre in the event of an unexpected crisis.

Income projections for support payments and grants

Support payments are vital for the survival of many farms. Ideally, these payments should be used to drive improvements in the business, but the reality is that the payments are a critical component in maintaining liquidity. Surviving between one support payment and another is a way of life for many as their arrival signals the opportunity to clear away critical debts. It is not unusual to find that many of the businesses and industries that support the farming sector will be carrying accounts receivable that is equivalent to over 300 days-of-sale. This is not only a great burden which, in being covered, will cause higher prices; it results in a financial ecosystem that is highly vulnerable to bankruptcies, which always have a wide and unexpected impact.

In the course of a year, the net cash flow from trading and support payments must be positive for longer term survival.

Grants usually come with commitments to spend and these monies must be kept separate in cash-flow terms. However, many grants are retrospective and only appear after programmes of work have been started or even completed. This is hard on cash management but it will avoid the traps of spending ahead of meeting commitments.

The primary cash flow for the business can now be projected. This comprises:

- operating balances
- scheduled (non-trading) payments
- cash flows (= operating balances plus scheduled payments).

Long-term finance

Borrowing, long term, in the form of loans should not be undertaken to cover operational deficits. Bank managers will be alert to such situations, but this is still a big problem. Loans are advanced only when banks are satisfied that interest payments will not be too great a burden and that in the case of default there is an underlying asset that could be realised. Getting this wrong leads to decapitalisation and when this starts it tends to become a continuous problem.

Mortgages will be secured on property and should be only used to secure ownership or to invest in facilities. However, many farm businesses have taken on mortgages to finance a transition from a failing business model to (prospectively) a more attractive one.

Long-term finance is a B/S item and the monies involved need to be accounted for differently. Its impact on cash flows comes when assets must be realised in order to settle trading debts. One of the objectives of producing cash-flow projections is to win sufficient time to realise assets when facing such an unhappy crisis.

Cover ratios

All finance charges must be met out of the free cash flow in a business. The free cash flow is simply the trading revenue less operational expenses. This is essentially the 2nd contribution (see Chapter 3). Ideally, the burden of finance charges, less any capital repayments, should be, at most, 25% of the free cash flow. This equates to a cover ratio of 4.

Businesses start to experience difficulties when its cover ratios fall below 2.

Many farm businesses have negotiated loans on the basis of their 3rd contribution, which is its cash flow after the payments of grants and support. This may well cause trouble if and when support payments are withdrawn.

Chapter 3
BUSINESS PERFORMANCE

3.1 THE HIERARCHY OF PROFITABILITY

When examining the profitability of a business, a hierarchy of tests should be carried out.

The **first test** of profitability is to establish whether a business makes a positive 1st contribution (sales revenue less variable costs). The 1st contribution is a measure of the primary cash flow in a business. If this is negative, then the business will be losing cash on every transaction and its true viability, if it is not in transition, will be questionable.

The authors have analysed over 320 farm accounts in recent years. When these accounts were examined closely it was found that 18% did not cover variable costs. These farm businesses continued to survive only because the injection of support payments, grants and the income derived from other family members covered the deficiencies.

The **second test** of profitability is to establish whether a business makes a positive 2nd contribution (1st contribution less fixed costs). If this is negative, the losses will have to be covered by injections of capital or the sale of business assets. Continued losses at this level will eventually decapitalise a business entirely and cause bankruptcy.

This test was failed by 80% of the farms in the study group. Again, support payments and other injections of cash kept the farms in business.

The **third test** of profitability, which is exclusive to farm businesses, is to establish whether a business makes a positive 3rd contribution (2nd contribution plus support payments and other income). As support payments are a central component of the farming sector in commercial terms, the 3rd contribution is currently the ultimate test. However, when businesses are managed at this level there are inevitable tendencies to *farm* the support mechanisms as much as the land or the livestock.

This test was failed by 20% of the farms in the study group. Support payments, and other contributions, had the effect of transforming the apparent profitability of farming from 80% being unprofitable on farming activities alone to 80% being profitable in the final analysis. As support payments come under political scrutiny, or are eliminated, **the ability to deliver a positive 2nd contribution becomes an imperative**.

3.2 ROTA

When a sum of money is deposited in a bank or a building society it can qualify for interest payments. When interest rates are set at, say 5%, a deposit of £100 will earn £5 in its first year, and leave £105 to attract interest in its second year. It is a simple concept (although the calculations are termed compound interest) and it neatly encapsulates the dual nature on money with respect to amount (value) and term.

This raises the question as to what would be an equivalent measure for a business? It is certainly not as simple as expressing profits as a percentage of sales and even this sort of measure would be confusing with a host of different possible measures of profit. The P&L of the statutory accounts can offer up PBIT, and PAT as candidates and the management accounts could offer three levels of contribution. Profit measures alone are further compromised by not reflecting the term aspects of money in a business. What about the sums on the B/S and what role should they play?

The total assets taken from a B/S represents the capital invested (deposited) in a business in all its various forms and, more precisely, it is a measure of the assets employed. The return on these assets will be some form of performance measure. Let us take a management (operational) perspective and select the 2nd contribution as a fair measure of profit. This produces the simple measure:

return on total assets employed (ROTA) = 2nd contribution / total assets employed

This nicely combines **amount**, from the P&L via the management accounts, with **term** from the B/S. It is the business equivalent of interest on deposits.

At this point, a mathematical trick can be introduced, which will improve the utility of this measure. By multiplying the ROTA parameter (2nd contribution/total assets employed), by sales turnover (above and below the dividing line), the parameter can be decomposed into two components (margin on sales and assets turn).

ROTA as a measure can therefore be decomposed into two elements; the profit margin and the assets turn (Figure 3.1). These secondary measures now reflect the value aspect of money (the profit margin) and the term aspect (the assets turn). The concept of **assets turn** is likely to be unfamiliar to many. Suppose a business has a sales turnover of £1,000,000 and its assets employed amount to £350,000. The asset turn is 2.86 (being 1,000,000/350,000). A better way to appreciate this is to turn the 2.86 into its days-of-sale equivalent. This would be 128 days-of-sale (being 365/2.86) and it signals that it will take 128 days for the sales to cover the assets employed in the business.

This now raises the question as to what would be an appropriate value for a ROTA objective? If interest rates are 5%, it could be argued that this is a datum for money free of risk. However, as businesses present very real risks it would not be unreasonable to argue that a premium of 10% could be justified over risk-free deposits. This would suggest a ROTA of 15%.

It should be noted that the assets turn dimension (Y-axis) is a consequence of changes to the B/S and the margins dimension (X-axis) is a consequence

Figure 3.1 The ROTA formula.

As defined above,

$$\text{ROTA} = \frac{\text{2nd contribution}}{\text{Total assets employed}}$$

This equation can be multiplied (top line and bottom line) by sales without affecting its value.

Then,

$$\text{ROTA} = \frac{\text{2nd contribution}}{\text{Sales}} \times \frac{\text{Sales}}{\text{Total assets employed}}$$

But,

$$\frac{\text{2nd contribution}}{\text{Sales}} = \text{Operating margin (X)}$$

and,

$$\frac{\text{Sales}}{\text{Total assets employed}} = \text{Assets turn (Y)}$$

Therefore,

$$\text{ROTA} = \text{Operating margin (X)} \times \text{Assets turn (Y)}$$

If ROTA is a constant K (as an objective), then,

$$X \times Y = K$$

This is a geometric conic section known as a hyperbola.

---

of changes to the P&L. This convention clearly separates the two different aspects of money, being amount and term, without confusion.

In Figure 3.2, the blue curve is the 15% ROTA line. Anywhere on this line will deliver a 15% ROTA but businesses will find some positions easier to achieve than others.

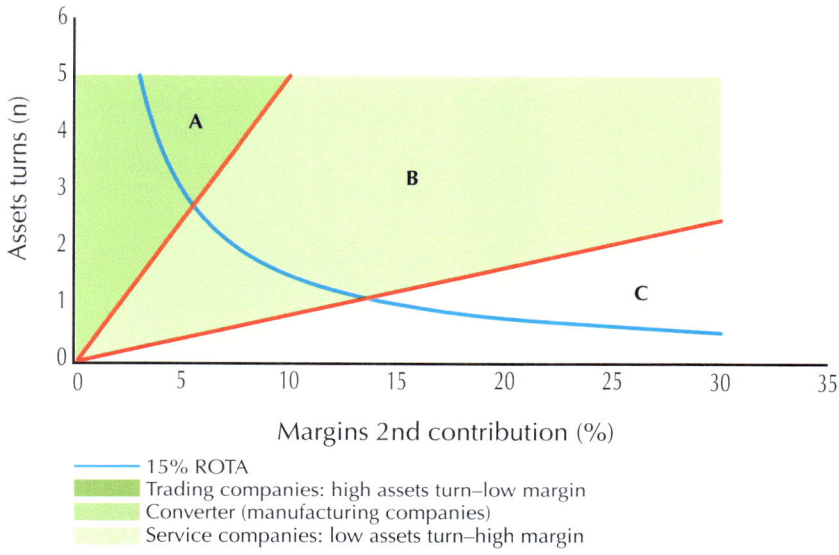

Figure 3.2 A 15% ROTA diagram.

In zone A, asset turns are high and margins are low. These are the characteristics of a trading business. A good example would be a company trading in precious metals where margins may be, say, 0.5% but assets turn could be 250 when books are cleared every working day. Such a business would produce a magnificent ROTA of 125% (equal to 0.5 × 250).

In contrast, in zone C asset turns are low and margins are high. These are the characteristics of a service business. A good example would be a firm of lawyers. Legal cases take a considerable time before they are concluded, and firms can wait a long time to be paid. If asset turns for a firm are, say, 0.5, reflecting the situation where accounts take 2 years to be settled, margins can be as high as 100%. Such a business would produce a ROTA of 50% (equal to 100 × 0.5).

Zone B is characteristic of converter businesses such as manufacturing industry. Such a business might deliver a margin of just 5% and experience an asset turn of 4 delivering a ROTA of 20% (equal to 5 × 4). Theoretically, farming is a converter business but the heavy assets base (in land and breeding livestock in particular), reduces the prospects for a good assets turn and farm businesses need good margins to compensate.

At this point you might ask, so what? How do I use a ROTA analysis?

Apart from measuring how your business performance compares with money on deposit in a bank (for which a diagram is hardly necessary) the diagram can be useful in two other important ways.

- ROTA results can be plotted over a number of years and the track that results will demand explanation as it moves around. In tracking the results for some farms, the impact of weather and other events become obvious. The authors have even seen instances of the tracks going in circles as farmers change back and forth between different crops or farming practices.
- The ROTA diagram can be used to set new performance targets to take a business from its present position to meet a new objective. The new objective point will quantify the improvements to be secured in terms of assets turn and margins.

For the farm represented by the set of management accounts set out in Figure 2.2, with assets employed at £850,000, the improvement in ROTA in going from a pre-support basis (the 2nd contribution) to a post-support basis (the 3rd contribution) is shown by the red line in Figure 3.3, indicating an improvement from −1.18% to 4.71% ROTA.

3.3 CASE STUDY: ROTA TRACKING

The ROTA performance on a livestock farm in northwest England was tracked over a 7 year period and the results demonstrate the utility of the ROTA analysis in business management (Figure 3.4).

The movement, represented by the red line with the blue line being the 15% ROTA objective, was highly volatile and demanded a logical explanation. Two of the years were badly affected by excessively wet weather, which limited practical grazing. These years are easily identified on the ROTA chart.

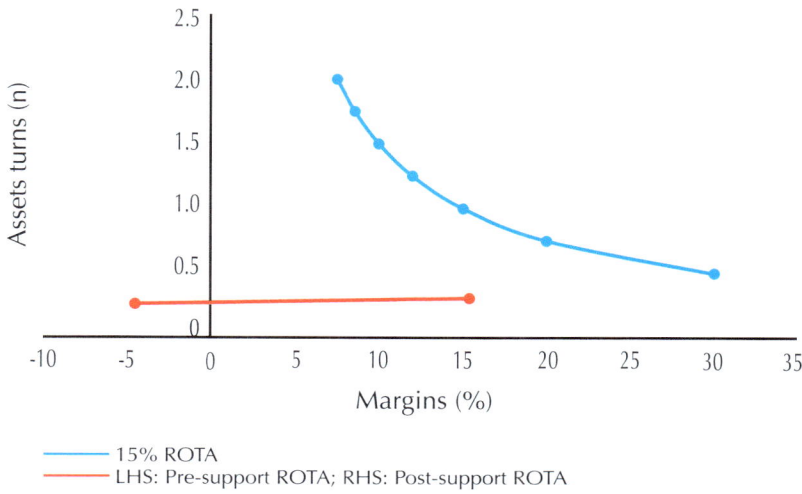

Figure 3.3 ROTA improvement: pre-support to post-support.

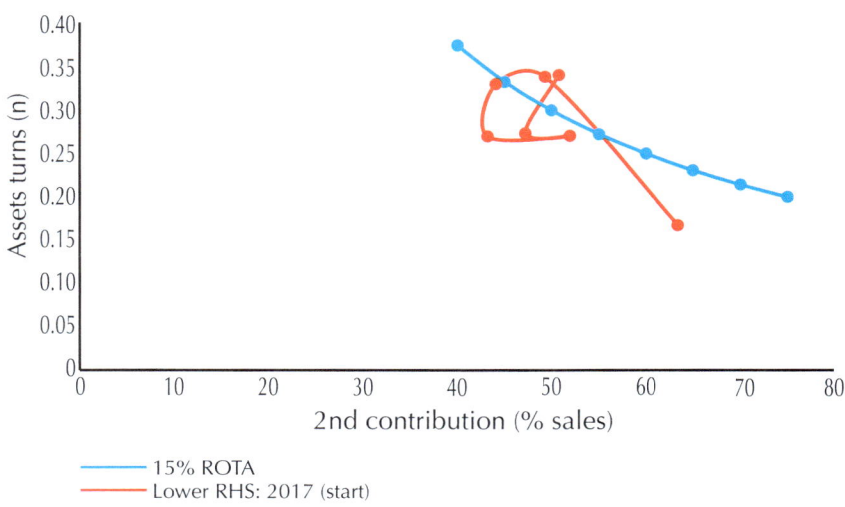

Figure 3.4 ROTA: 7 year track.

Other wild movements were explained by a move into a specialised breed for three years and a subsequent change of mind and a reversal of policy.

3.4 PRODUCTIVITY

Productivity in farming will come from one of four sources.

Capital productivity is primarily related to the effectiveness of an investment in changing the landscape. This would encompass, for example, spending on programmes to create drains or construct fences and walls. A secondary case would relate to an investment in some form of factory-type processing such as a dairy parlour, cheesemaking or butchery. The productivity improvement that results from such expenditures is in the form of cost reductions over the former situation. Capital, however, is easily squandered and capital productivity can be difficult to achieve, especially when added-value is confused with added-cost.

> ### CAPITAL PRODUCTIVITY SHOULD BE MEASURED BY EVALUATING THE RETURN ON INVESTMENT
>
> Consider the case where a dairy farm wishes to invest in a new covered high-capacity slurry pit, which is budgeted to cost £200,000. With the ROTA test rate of 15% for a farm business, and if the slurry pit is assigned a life of 15 years, the operational savings should deliver an on-going £30,000 per annum. If the money had to be borrowed, the finance charges, at 5% interest, with capital repayments would be £23,333 per annum. The difference, £6,667 per annum, can be considered to be a gain for the economic life of the slurry pit.

Mechanical productivity is related to the use of plant and machinery to deliver greater levels of output. In this regard, the tractor has been one of the most important and successful examples of mechanical productivity in farming. Some types of plant and machinery, such as balers or harvesters, that perform very specific tasks can suffer from very low degrees of utilisation as usage is invariably confined to just a few weeks every year. Although such equipment can be mechanically productive its cost-effectiveness can be very disappointing.

MECHANICAL PRODUCTIVITY SHOULD BE MEASURED BY REDUCTIONS IN THE UNIT COSTS OF PRODUCTION

Consider the case of a new automated vegetable picking machine, costing £60,000. Its role is to harvest 120,000 tonnes of vegetables. If it has a useful life of 10 years and requires annual maintenance protocols that are equivalent to 5% of its original capital value it should be charged out to the farm business at £9,000 per annum (£6,000 + £3,000). If the machine is operated by labour, rated at £30,000 pa for just three months (£7,500), its unit cost for vegetable picking is £0.1375 per tonne.

$9,000 + 7,500 = 16,500 / 120,000 = 0.1375$

If the original process involved 10 labourers, rated at £25,000 pa working for four months (£8,333), the unit cost would have been £0.6944 per tonne.

$10 \times 8,333 = 83,333 / 120,000 = 0.6944$

Mechanical productivity, provided the equipment can be used intensively to good effect, is one of the more attractive options on a farm.

Chemical productivity is related to the use of artificial supplies such as fertilisers and food concentrates. Such purchases can improve yields, at least in the short term, but the gains may not be commercially viable, and their use may have detrimental environmental consequences.

CHEMICAL PRODUCTIVITY IS OFTEN MEASURED IN IMPROVEMENTS TO YIELD

If an unfertilised field produces 4.50 tonnes of cereal per hectare but the field when fertilised by a proprietary chemical produces 7.50 tonnes per hectare there is a considerable increase in yield.

However, chemical productivity should be measured in financial terms, as with other productivity measures.

If cereal prices are £180 per tonne there is a revenue gain of £540 per ha. However, if the fertiliser use is 1.75 tonnes per hectare and prices are £350 per tonne the extra cost of production is £612.50 pa. This gives a very different picture and typifies the difference between out-put-driven perspectives and profitability-driven perspectives. This is an example of the difference between added-costs and added-values (see later), which highlights a productivity trap.

Figure 3.5a Capital productivity: fencing.
Credits: Alamy D8G5HP.

Biological productivity is related to investments in selective breeding or crop rotation regimes. The development of native breeds of livestock that can flourish on an all-grass diet and be left outside all year round has been a great achievement, for example. These initiatives can be very rewarding but can also be both unpredictable and deliver unintended consequences.

Some breeds of sheep, such as Swaledales, have been bred to produce in excess of 80 twins per hundred today whereas it would have been just 5 twins per hundred 50 years ago. This would suggest a huge improvement in yields but, while the births have increased, the ability of the ewes to produce a commensurate amount of milk has not. The consequence is that Swaledale lambs now require food supplements to be purchased to make up for the deficiency.

Figure 3.5b Biological productivity: Swaledale sheep.
Credits: AdobeStock_426379608.

Productivity measures on most farms tend to be related to some technical aspect of performance such as **annual dry matter production (kg/ha)**. While this is something to maximise, it should not imply it ought to be maximised at any cost. These measures sit quite comfortably with many farmers but fall short when little account is taken of the costs involved in pursuing such targets. However, the tacit assumption made by many farmers is that better yields must inevitably lead to a situation that is better all-round. This perspective must change if farms are to be successful commercially.

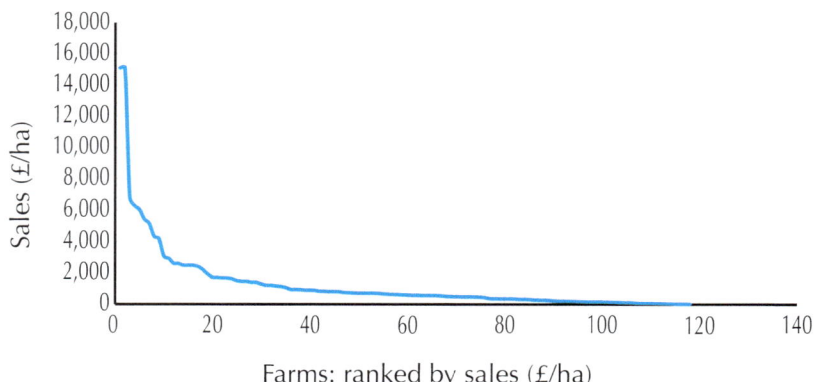

Figure 3.6 Farm productivity: a typical distribution.
Credit: Nethergill Associates Database.

Productivity, in a commercial sense, is a measure of the ability to generate sales. On a farm, a simple and effective measure would be **£ sales/ha** or **£ sales/LSU** where an LSU is a common livestock unit. A typical profile for a small group of farms can exhibit a wide range of values which reflect a wide variety of circumstances.

3.5 PROFITABILITY

Profitability is a measure of the ability to generate profits. On a farm a simple and effective measure would be **2nd contribution/ha (£)** or **2nd contribu- tion/LSU (£)**. The patterns of profitability for a group of farms can be very different from their corresponding patterns of productivity. There will be no direct relationships as the circumstances behind each measure will be very different (Figure 3.7).

The simple pursuit of output, above all else, is most unlikely to be reflected in a corresponding profit. However, the objective of a business enterprise is to be as profitable as possible. When the corresponding measures of pro- ductivity and the profitability on farms is analysed, the results can be sur- prising (Figure 3.8).

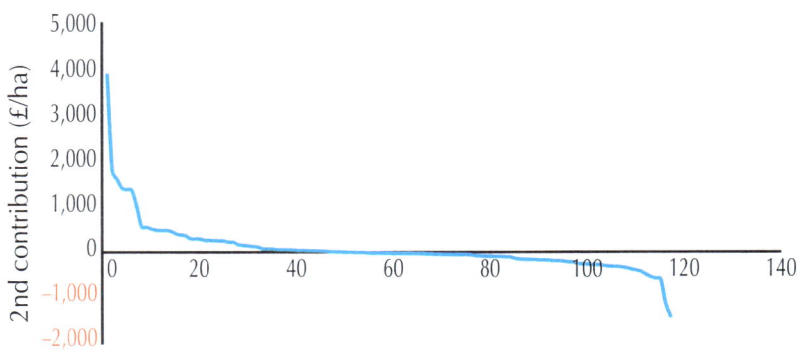

Farms: ranked by 2nd contribution (£/ha)

Figure 3.7 Farm profitability: a typical distribution.
Credit: Nethergill Associates Database.

This raises a critical question. Why are some of the more productive farms such poor performers in profitability terms? The reasons can lie in the confusion between added-cost and added-value.

3.6 ADDED-COSTS AND ADDED-VALUE

Added-costs are incurred whenever practices comprise activities that are not the least-cost options, consistent with the delivery of quality products in a responsible way. Such costs, which often reflect vanities, are notoriously difficult to recover commercially as they rarely command a sufficient price premium. There are numerous instances to be found, for example, of farmers working with non-native breeds that cannot be sustained on all-grass diets and are unsuitable for year-round outside rearing. Despite the often-great commitment and attachment to these breeds, the added-cost involved often outweighs the added-value in that the price premium is never sufficient to be commercially attractive.

The examples examined in the reviews of chemical and biological productivity (see above) well-illustrates the problems when added-costs outpace added-values.

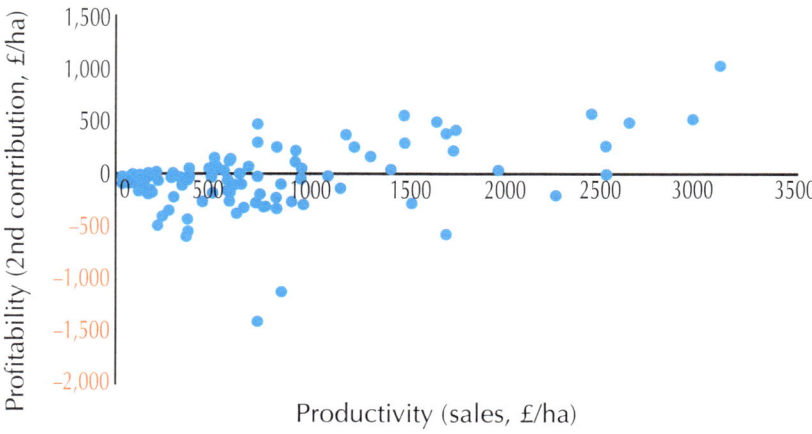

Figure 3.8 Profitability v. productivity.
Credit: Nethergill Associates Database.

Added-value, however, is more than just maximising the price premium. The production of milk, cheese and meat products would represent a typical range of added-value activities to be found on some farms. These activities offer a prospect for greater profitability by putting a greater work-content into the farm produce; by being able to market these goods as being well-differentiated from others; and by then exploiting the subsequent benefits of branding.

Added-value activities will require additional investment and will demand a marketing mentality. This brings new business risks in the guise of possible failures to produce sufficient cash flows to justify the investment and the likelihood that as activities become more specialised, they become more vulnerable to changes in the marketplace.

To make a success of added-value initiatives, transfer pricing must be understood and rigorously applied. Too many added-value business initiatives, which have been inspired by a need to make a farm business more profitable, have retained a commodity-business mentality when marketing the output of the new investment. Taking a sounding on market prices may seem logical in a competitive market but this can lead to serious under-pricing. When produce is underpriced, it will generally sell well but it will never deliver enough value to cover for an unprofitable farm business in the absence of any transfer price. Farm output should never be on a 'free-issue' basis to an added-value operation.

Consider a specialist cheesemaking business on a dairy farm. When milk prices were £0.38/l the operating costs of the farm were £120,000. When a transfer price premium of 25% was added to these costs, to total £150,000, the 130,000 l of milk produced should have commanded a transfer price to the cheesemaking business of £1.15/l. While this was a considerable advance on the 'market price' of £0.38/l the cheesemaking business could still produce a competitive cheese. The customers valued the cheese and met its premium price and the farm was fully rewarded for its role in the supply chain. When the business had no transfer price the cheese failed to realise its true value and the farm was burdened with losses at the 2nd contribution level as output costs were equivalent to £0.92/l and the selling price for milk was £0.38/l.

From the Nethergill Associates Database, the average 2nd contributions, expressed as a percentage of sales for dairy farms, lowland livestock farms and upland livestock farms were 12.99%, –11.07% and –40.48%, respectively. This demonstrates that the added-value activities of the dairy farms delivered better results than the lowland livestock farms that were selling on a commodity basis to abattoirs. The upland livestock farms were also selling on a commodity basis to abattoirs with products undifferentiated from their lowland counterparts. However, their results are significantly inferior reflecting the added-costs of correcting for their disadvantages in elevation, among other things.

Not everyone can be the least-cost producer in a marketplace; it is by definition only open to a single supplier. Other producers can evaluate the competitiveness of their businesses through a process of benchmarking (see 3.8).

3.7 THE MARKET TRAP

Many farm businesses have diversified so as to offset the losses made on the farm. This is a false objective as it does nothing to address any unprofitable activities on the farm. The objective of an added-value or downstream business should be to take farm outputs at cost plus a margin and set the selling price of the added-value produce accordingly. This does not prevent efforts being put into reducing farm costs when there are opportunities. The logic for this approach is that the added-value product can command a premium price if the product is differentiated (and so recover the actual costs of production) whereas farm outputs, with no added-value, will in all probability be constrained by prevailing commodity prices and have little scope for premium pricing irrespective of any intrinsic qualities.

There are two traps to which businesses can succumb. First, when it is argued that the added-value product has a 'market price', which should prevail. This simply guarantees that the farm element of the business will continue to run at a loss. Second, as often happens in cheesemaking, it is argued that the source commodity, milk here, has a 'market price'. In this case the essential quality in the milk for cheesemaking, which will make it

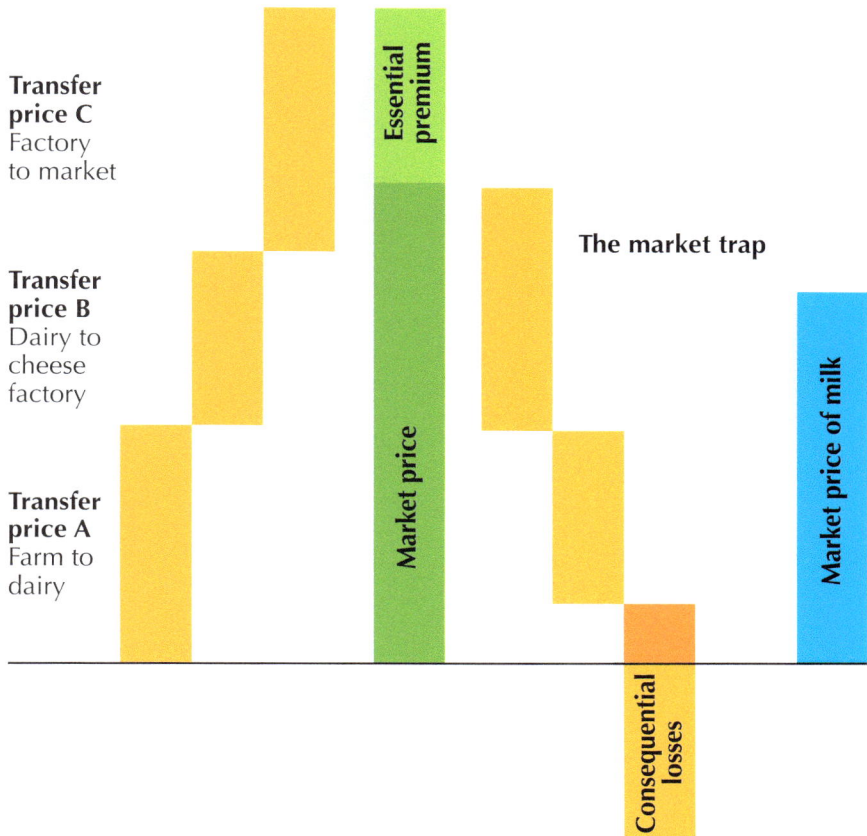

Figure 3.9 An illustration of a market trap: cheesemaking.
Credit: Nethergill Associates. Illustration: Elaine Leggett.

more expensive than to produce that milk for supermarket shelves, is not being fully recognised. Again, this simply guarantees losses will be sustained on the farm (Figure 3.9).

3.8 BENCHMARKING

Benchmarking is the practice of using industry performance data for the purposes of comparison in the quest for improvement, either physically or commercially. There are three quite distinct forms to consider.

The **industry sector benchmark** that slots business performances, often into deciles (one-tenths of the datasets) and provides a ranking within an industry sector or market sector.

> For example, in a group of 50 farms performance was measured by the 2nd contribution/ha and it ranged from £1,100/ha to £50/ha. A farm returning, say £500/ha can now be ranked within the group. In this instance the farm would have been 15th (in the 4th decile). This is a simple and often useful measure. If a farm business came 3rd (for example) it may well take some pride in knowing that on some aspect of performance it was in the top-rated decile. It is a measure of relative performance.

The **best-practice benchmark** is altogether different.

> For example, we know that is is unlikely that a farm will make a positive 2nd contribution if its fixed costs are greater than 40% of pre-support revenues. We also know that variable costs on a farm should always be in the range of 5–15% of pre-support revenues if the farm is operating at maximum sustainable output (MSO). So, we could set these as two benchmarks for a farm business. Performances are measured by indexing actuals against the best-practice parameter rated at 100. This is very effective for business planning. As it is a measure of absolute performance it will be much more demanding to follow or exceed.

The adoption of **standards based on the measurement of work-content**. This is an industrial concept and could, at some point, be applied to farming albeit at a level of detail that would be much less precise than that required by industry.

> For example, in industry a standard might be set for the time to turn a metal seal of a given size on a given lathe. This would be measured in

secs/unit and the time needed to complete any batch of work is then simply found by multiplying the quantity by the standard. In farming such detail would be impossible for an activity like harvesting crops as no two farms can ever be the same in terms of weather, access, shape, dimensions and soil type.

The actual values of these standards are un-important; it is the methodology that establishes them that is critical. These standards are not used for comparative purposes; they are used to simplify the computation of profitability through a mechanism called variance analysis. Their primary role is to ensure that activities are proceeding according to plan (budgets).

While there is a wide variety of benchmarking datasets available, usually compiled by industry groups or interest groups, most of them will be of the industry sector type and will focus on physical performance measures relating to usages and yields. Few will have financial or commercial benchmarks.

Compared to manufacturing industry, the farming sector is not well-served by benchmarking services offering reputable methodologies and proper statistical levels of significance. To be fair, industrial standards and benchmarking in manufacturing did not appear overnight. They have been developed from a long and distinguished contribution by industrial engineering sciences in a way that, currently, has no equivalent in farming.

3.9 BUDGETING

Advanced manufacturing industry has used standard costing practices for over 100 years. It has used these standards to great effect. Their origins are to be found in the pioneering work in industrial engineering sciences driven by the great manufacturing industries of the mid-west of the United States in the years following the First World War.

A business will evaluate its operations at standard and then track the variances to its standard projections. The net variance is the corrective factor on the profits projection at standard. It is a process of management by exception, in that a standard profit is assumed and the net variances will adjust this to actual performance (see Chapter 4).

Farming is a very long way from the prospect of industrial quality standard costing models but the concept of management by exception has some validity.

Budgeting, with a strong emphasis on the management of cash flows, is an essential exercise in farming but it is largely a neglected practice. The format for a farm budget should follow that of the management accounts with additional lines below the trading cash-flow summary to account for movements of capital to and from the B/S, and the servicing and repayments of loans. This is explored more fully in Chapter 8.

CHAPTER 4
ECONOMIC MODELS

4.1 THE STANDARD THEORY OF THE FIRM

The standard model was the result of attempts in business to establish an economic **break-even point** for an enterprise – the point in volume terms, beyond which, profits will be inevitable. The model comprises three essential components:

- revenues related to output volumes [R]
- fixed costs (for the business enterprise) [F]
- variable costs related to output volumes [V]

Break-even occurs, on this model, when:

Revenue [R] = Fixed costs [F] + Variable costs [V]

This is best shown diagrammatically.

In Figure 4.1 the yellow line represents the production revenues as volumes increase; the red line represents the fixed costs (which are incurred irrespective of any output); and the teal line represents the costs of production as volumes grow. The break-even point is the place where the yellow and teal lines intersect.

As support payments feature heavily in farm businesses, the blue line represents the total revenues (post-support payments) as volumes increase and

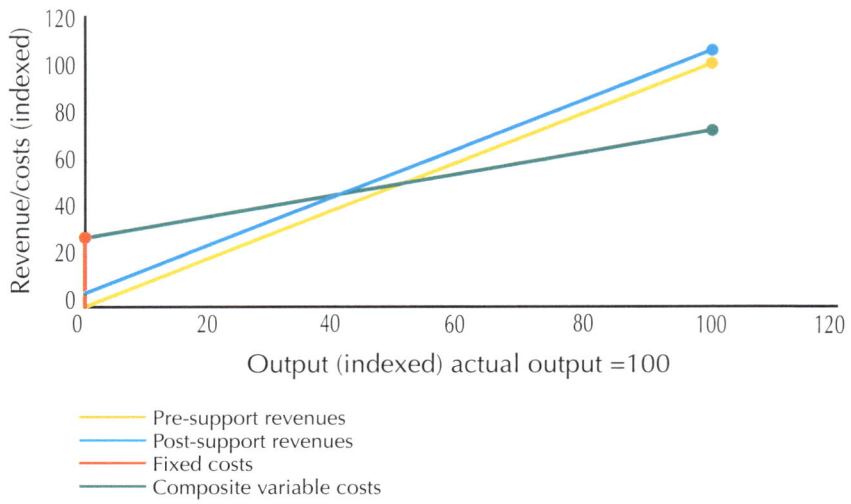

Figure 4.1 Break-even analysis: the standard model of the firm.

this gives an alternative break-even point at the place where the teal and blue lines intersect. It can be seen that this point comes before the pre-support break-even point and will always do so.

The underlying assumption behind this model is that variable costs are directly related to outputs – as output grows so will variable costs in some fixed proportion. This is referred to as a **linear relationship**. A non-linear relationship would be some kind of curve, as opposed to a straight line.

This model has been used extensively by manufacturing industry with great success. It is simple and accurate in these applications. The linear relationship model has been offered extensively by farm colleges and the implications of driving outputs will be familiar to many farmers. The model suggests that profits grow, inexorably, beyond the break-even point in a farm business and that **the pursuit of volume can only be good in a commercial sense**.

With such a simple and unequivocal message farmers can be forgiven for thinking that additional profits improvement will always come from the relentless pursuit of increased yields and the containment of fixed costs.

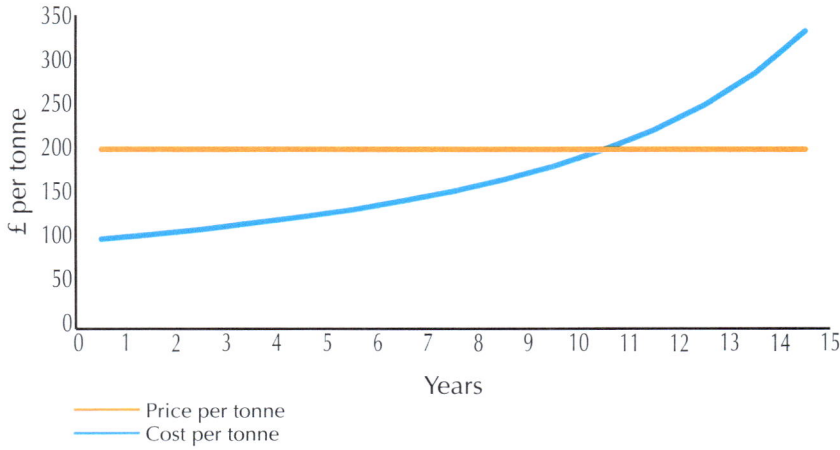

Figure 4.2 Example of non-linear variable costs: mining a mineral ore body.

Variations of this model have been applied to cases where the variable costs are non-linear. The mathematics are a little more complicated in these cases, but the essential principles are the same. An example non-linear case in industry would be, say, a mining operation where cost rates increase as the ore body is depleted and extraction becomes more difficult. This is shown in Figure 4.2.

However, farming turns out to be very different in practice from the assumptions regarding the behaviour of costs implied in the standard model.

4.2 FARM VERSUS FACTORY: A CONTRAST

The factory system is based on three inter-related concepts.

- The principle of the division of labour: productivity is increased when people are specialised in a task (*execution*) and work, sequentially, in a highly repetitive fashion.
- The principle of taking work to the workers in a common establishment: production is concentrated and not dispersed.
- The principle of the economies of scale: scaling-up production with the aid of evermore-productive machine tools offers the opportunity

to reduce unit costs. All productive equipment in a factory will tend to be run at maximum levels of utilisation.

This is diametrically opposite to the circumstances prevailing in farming.

- Farmers are generalists, not specialists, and undertake *complete tasks* incorporating elements such as planning, execution and control. Whereas a factory worker may have a repetitive job with, say, a 3-minute work-content before repetition, a farmer will have daily jobs modified by the seasons.
- Farmers have to work in the field wherever the jobs are.
- Farmers work with Nature, which cannot be scaled-up like some machine tool. The productivity of Nature will have absolute limits. Additionally, productive equipment on a farm will tend to have low levels of utilisation, being influenced heavily by the seasonal nature of the work.

Translating industrial practices directly into farming is fraught with danger.

4.3 UNDERSTANDING VARIABLE COSTS

Uniquely, variable costs in farming decompose into two different classes: productive variable costs (PVCs) and corrective variable costs (CVCs) (see Figure A.6 in the Appendix). This is driven by energy considerations that are unique to farming. Additionally, to add further complications, the two types of cost are incurred sequentially, not concurrently. If the costs were concurrent, a simple non-linear relationship would be the result.

PVCs are those costs on a farm associated exclusively with the use of natural resources in the production process. The key natural resources on a farm are sunlight, precipitation, its soil fertility and its underlying geology. Activities related to these natural resources include planting, harvesting and animal husbandry. These natural resources are a source of 'free-issue' energy.

CVCs are all those costs on a farm associated with substituting for an absence of natural resources or offsetting some of the disadvantages of geography. These costs include purchased items such as fertilisers, herbicides, pesticides, and food supplements and concentrates. CVCs are also incurred when attempts are made to offset some of the geographic disadvantages of elevation, precipitation and latitude. In such cases there is a tendency to force the landscape, for example by the extensive use of artificial fertiliser, so as to compete with yields available on farms with fewer geographic challenges. However, the critical distinction between PVCs and CVCs is that CVCs will always have an essential 'industrial energy' content.

Logically, CVCs should only be incurred after all the available natural resources have been exhausted, hence the sequential nature of the costs structure. It is the recognition of these differences that invalidates the standard model of the firm in farming applications and transforms the underlying economics of farming.

4.4 UNDERSTANDING FIXED COSTS

Fixed costs are essentially related to running an enterprise and will be incurred irrespective of the output produced. These costs, too, can be decomposed into two types; the unavoidable (and necessary) costs that fund essential assets and administrative capabilities and those lifestyle-related costs that often get entangled in business ventures.

Managing fixed costs is a much simpler challenge (intellectually) than managing variable costs. The objective should be to keep these costs to a minimum and eliminate all lifestyle-related costs from the business.

The biggest challenge will be to contain the costs associated with plant and equipment that can only return low utilisations.

4.5 STANDARD COSTS

The manufacturing industry has developed models for standard product costing to simplify the business of exercising cost control and making profit projections quickly. These models involve the detailed specification of manufacturing methods, production rates on machine tools, labour inputs and costs, and consumable supplies and costs. The typical environment for manufacturing industry is one where most of the factors in production are easily measurable if care is taken. Factors outside direct management control tend to be very limited (typically only prices for supplies) once operations are under way.

In contrast, a farm is exposed to factors with huge levels of unpredictability, such as the weather and commodity market prices. Consequently, farming has had little exposure to standard costing initiatives and the field of industrial engineering in general.

4.6 THE THEORY OF MARGINAL COSTING

Suppose a dairy business is producing 1 million litres of milk, at a £0.35/l overall, for a price of £0.40/l on a comparable basis. At some point the farmer might reasonably ask the question: where would I be if I decided to expand production by, say, 100,000 l per annum? If the price stays the same, at £0.40/l, the cost of the new and additional 100,000 l must be less than £0.40/l to make commercial sense. This is the concept of marginal costing.

To be understood properly, it must be appreciated that the costs of producing the original 1 million litres must be kept separate from the additional and new 100,000 l. If the cost of the additional milk production is £0.45/l the expansion makes no commercial sense. However, if the two situations are combined the average production cost will be £0.36/l overall and that the business is still profitable. This is the trap in marginal cost analyses when it is not properly applied.

When it is applied properly, it works for cost profiles that are linear (straight line) and non-linear (curved-line), provided the costs behave in one continuous fashion. A linear cost profile will exhibit a break-even point, beyond which further expansion becomes increasingly more profitable up to the point where capacity runs out. A non-linear situation will also exhibit a break-even point but, if costs increase in general as output grows (as would be the case in mining a limited body of mineral ores), it might also break-back into unprofitability. In these non-linear situations, there will be a point of maximum profitability (which is found by using differential calculus). The marginal cost for a linear cost profile will be constant; for a non-linear profile it will change with output and, typically, increase.

Some cost profiles are sequential, and consequently non-linear (even though the sequential components may comprise linear profiles). This happens often when further expansions need new investments, for example. In these cases, marginal cost analyses can be tricky. To avoid the traps of marginal cost analyses, unit cost profiling should be used as an alternative. This will be dealt with later in Section 4.8

4.7 THE MSO MODEL OF THE FARM

In farming, we have established, above, that variable costs decompose into two sequential components, the PVCs related to working with Nature, and the CVCs related to substituting for Nature. This reality changes, fundamentally, the shape of the break-even analysis. There will now be a clear position of maximum profitability, together with a break-even point that will be quite different from that suggested by the standard model and a probable break-back point, too (Figure 4.3).

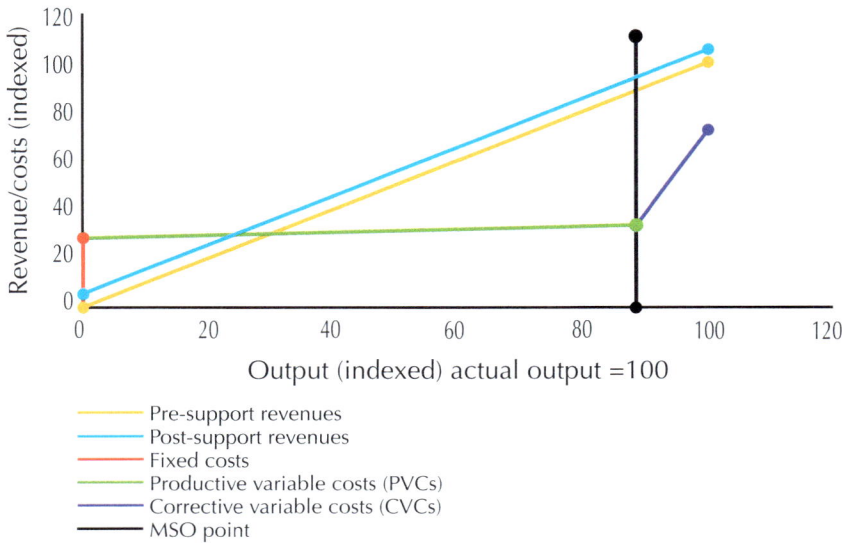

Figure 4.3 Break-even analysis: the MSO model of the farm.

On a typical farm, the PVCs increase at a lower (slower) rate than pre-support revenues. If this is not the case a farm will lose cash on every transaction and will be fundamentally unviable in the long term. However, almost invariably, CVCs increase at a higher (faster) rate than pre-support revenues. This is a reflection of two phenomena: the embedded energy content of substitutes can never be fully recovered in the food produced and the effect of the high-cost industrial energy content of supplies having to be recovered in the intrinsically low-value energy content of food products. When all the free-issue energy resources available on a farm have been used, the inflexion point, between the last of the PVCs and the onset of the CVCs, is termed the point of **maximum sustainable output** (MSO).

At MSO farm profits are maximised. The sustainable aspect is attributed to the fact that, with an absence of CVCs, the managed landscape will have no human adulteration to affect its long-term and permanent fertility. Farms with a long-term practice of heavy fertiliser use tend to evolve, or be changed by design, into a pattern of *monoculture*. In arable farming, this is done to force-up yields, but beyond a certain point yields tend to stabilise and greater inputs of agri-chemicals are needed to maintain outputs.

In Germany, the trade association of fertiliser manufacturers, NABU, have conceded that wheat yields grew continuously from 1961 to 2000 (at c.2.3% per annum)[2] but have flattened since then. NABU attributes this, dubiously (the authors contend), to climate change. There are, today, some highly intensive arable farmers that speak in terms of projecting only a further 80, or so, harvests to sterility unless the farming model changes. While this is contested in many quarters[3] it is reflective of the situation where farms are using increasing amounts of artificial fertilisers (CVCs) to maintain levels of output. Certainly, some farmers who are reducing their CVCs (implementing MSO programmes as a client of Nethergill Associates) have reported to us that there has been a noticeable increase in the depth of their soil.

The MSO point coincides with being:

- the point of its *maximum* profitability
- the point of its *minimum* energy footprint (see Section 5.6)
- a point of *optimised* biodiversity.

The energy footprint is minimised in the sense that no inputs containing an industrial energy content are used in production and that all the free-issue energy resources available have been used to best-effect. The pattern of biodiversity is optimised in the sense that it is simply uncompromised by any farming inputs and simultaneously delivers the best commercial outcome for the farm; this pattern of biodiversity may still, however, disappoint some constituencies of opinion among environmentalists and conservationists. Changing the patterns to some prescription will not only incur more expenditure but will never deliver a predictable result.

Most farms in the UK are operating above MSO. However, from the Nethergill Associates Database, up to 10% of farms appear to be operating below MSO, in that not all the free-issue energy available is being used. These farms will also experience an inflexion point but after eliminating its CVCs it will have an opportunity to expand to its MSO point.

As farms have pursued a programme to eliminate CVCs evidence suggests that the MSO point tends to increase as soil fertility improves.

4.8 UNIT COSTS

Businesses comprise a series of fixed and variable costs.

If the fixed costs in a business are, say, £100,000 and the variable costs are of a linear form and represent £5 per unit of output, then consider the difference in unit costs when 5,000 units are produced to that when 10,000 units are produced. The situation is summarised in Figure 4.4.

The unit costs will be seen to reduce from £25/unit to £15/unit. As volumes increase further, the unit costs will reduce continuously, but at a decreasing rate. This is shown in Figure 4.5.

In farming an MSO model should be applied.

Figure 4.4 Unit cost calculations.

Production	units	5,000	10,000
Fixed costs	£	100,000	100,000
Unit fixed costs	£/unit	20.00	10.00
Unit variable costs	£/unit	5.00	5.00
Total unit cost	£/unit	25.00	15.00

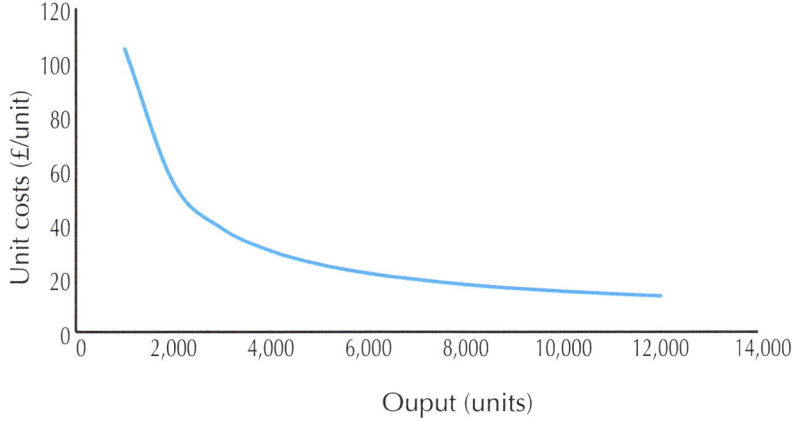

Figure 4.5 Unit cost profile: the standard model of the firm.

If the fixed costs of a cattle livestock business is £20,000 and the PVCs are £300 per beast, up to 30 beasts, and £1,200 per beast afterwards up to 40 beasts in total (as a consequence of CVCs being incurred), then consider the difference in unit costs for the 30th beast and that for the 40th beast.

The unit cost for the 30th beast is £967 and £1,700 for the 40th beast. If cattle prices are £1,100 per beast at slaughter, losses will be made on any cattle over 30 in number, as each beast costs £1,200 in variable costs at this point before any fixed costs are added.

Notice, however, the marginal cost trap: The total cost incurred by the business is £41,000 to produce 40 beasts and this suggests that the unit cost at 40 beasts is £1,025 and, therefore, all is well with prices at £1,100. It should be noted that although the business is profitable with 40 beasts (at £3,000) the business is even more profitable with 30 beasts (£3,990).

These two cases, the standard model and the MSO model, have very different unit cost profiles. These are compared in Figures 4.6 and 4.7. Figure 4.6 examines the general case in moving from the standard model shown in Figure 4.5 to its corresponding MSO model. Figure 4.7 examines the specific case and is set out in a form which would be produced for a farm business.

The orange profile represents the standard model, as set out in Figure 4.5. The blue profile represents the MSO model with its characteristic division between PVCs and CVCs. This line reaches its unit cost minimum at the MSO volume of output. The orange and blue profiles cross-over (dotted red line) at actual levels of output, as the total costs at this point will be the same for both models. On the standard model it would appear that unit costs continue to fall for outputs above actual levels while the reality is that unit costs will have been increasing since passing the MSO point.

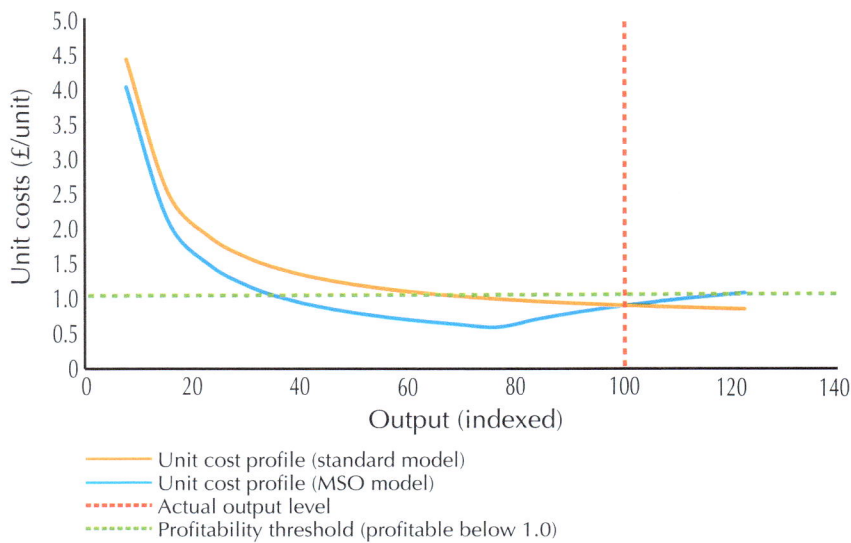

Figure 4.6 Unit cost profiles: standard model v. MSO model.

The chart in Figure 4.7 has been configured in an unconventional way to help with avoiding the traps of marginal costing. The unit costs have been expressed as £/unit per unit (this is a process of normalisation so as to produce a standard format and standard parameters for all farm businesses). That is, for profiles that dip below 1.00 £/unit per unit, extra output will be worthwhile. This is shown by the dotted green line. Marginal costing is then a simple case of finding the value of the blue line, at a particular output, and if it is below the dotted green line then extra output is worthwhile.

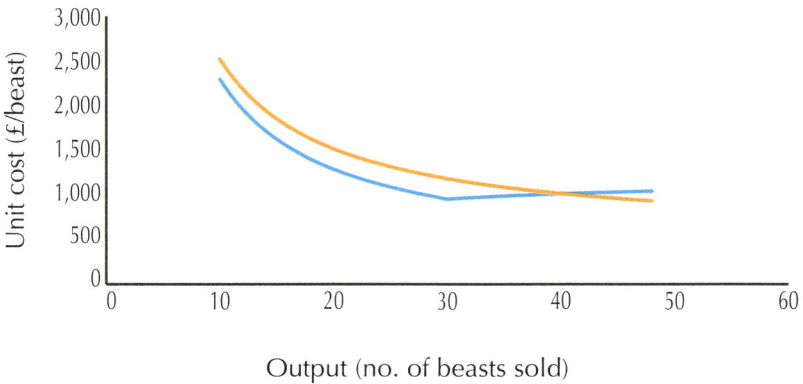

Figure 4.7 Unit costs: standard model v. MSO model.

CHAPTER 5
ENERGY

5.1 PHYSIOCRACY

There are two diametrically opposed concepts in society at large about what should comprise the real economic value of Nature and the environment. One concept argues for a quantitative stance and attempts to ascribe a monetary value to natural resources; the other concept argues for qualitative stance and an 'enjoyment of life' valuation.

While this may seem to be a recent phenomenon, associated with a more environmentally conscious age, the concepts actually have a long pedigree. In 1758, the French economist Francois Quesnel led a school of economists that believed that natural resources were the source of all material wealth. He coined the term physiocracy (literally, *the rule of nature*) and he and his fellow physiocrats maintained that the economic process could only be exploited by focusing on a single issue: the productivity of agriculture.

The physiocrats proposed that the economic process was subject to certain objective forces that operated independently of human free-will, which they called Natural Law. They argued that Natural Law had two components.

- Physical Law: activities advantageous to society.
- Moral Law: the sentiments driven by the consequences of Physical Law.

Their timing was unfortunate for history, as firstly, the nascent industrial revolution attracted the immediate attentions of economists, and, secondly, the science behind their thinking did not emerge until Rudolf Clausius and Ludwig Boltzmann worked on the physics of thermodynamics after 1865.

5.2 THERMODYNAMICS

Food is a fuel and all fuels are a form of energy. Farming, as a consequence, can be regarded as part of the energy sector of the economy and it will be constrained by the same physical realities that all energy producers will encounter. There are four inescapable **laws of thermodynamics**, which underpin all considerations of energy. These can be stated, in a form relevant to farming, as follows.

> **Rule A** When farming comes into contact with the natural environment, Nature will find a position of common equilibrium to accommodate a mutual coexistence.

This may take some time. Most importantly, this process has resulted in our **managed landscape**, which has become, over the centuries, our effective natural landscape. Technically, the managed landscape is in a state of *dynamic equilibrium* with Nature in that a small amount of work has to be put into it to maintain the status quo.

> **Rule B** All and any changes to the landscape will absorb or require work (and hence, energy).

No change comes free. This will come as a surprise to those advocates of rewilding by the simple expedient of doing nothing. When natural energy, in the form of sunlight for example, is declined by the economy there is a cost associated with the lost opportunity.

> **Rule C** The work involved in producing change in the landscape will be irrecoverable. The increased level of order in a changed landscape

will always be less than the increased level of disorder left behind in the world at large.

This is the notorious 2nd law of thermodynamics in physics. Essentially, the managed landscape represents an accumulated investment in productive assets by society. Failure to maintain this asset will deplete the value of the natural capital within the economy.

However, to understand the 2nd law more fully the concept of **entropy** must be understood (see Section 5.3). Everything in life is subject to a process of decay and the effects of, what physicists term, the **arrow of time.** Decay is a process that takes something with a degree of order and then makes it progressively more disordered. When this happens the entropy of a decaying system is said to increase.

It now might seem that evolution, and economic progress, which are the antitheses of decay become difficult to explain. Evolution, which is a change from less-ordered forms to more-ordered forms, can only happen when profligate amounts of energy are consumed to make this happen and it is this energy that is dissipated and not recoverable.

Rule D There is no possible pathway back to the original wilderness.

Failure to maintain the managed landscape will result in a new, and unpredictable, form of wilderness not some former landscape. The future for any landscape is a consequence of the conditions prevailing at the point when changes are imposed either by action or by default. As Nature, when left alone, will change by a series of random choices new landscapes will be quite unpredictable. One of the authors attempted to create an area on his farm that should have attracted more lapwings; it failed in this, but it did attract new colonies of red squirrels.

When environmental schemes fail to meet such expectations there is a temptation to intervene and bias changes into some more desired form. This is little short of catastrophic in terms of wasteful energy consumption.

5.3 ENTROPY

Entropy is a concept that often bewilders many. Yet, in essence it is very simple. Its strange outcomes can be baffling. Everything is supposed to be in a state of decay, but the managed landscape and the economy in general appears to be quite well-ordered. How can this be reconciled?

If we start with the situation that at some point in time Nature comprised a degree of disorder A and some lesser degree of order B. To change this, according to Rule B, work has to be expended. The energy to provide this work will come from the sun. After some time, Nature will have changed to a new set of circumstances, whereby the degree of disorder has increased to X and the relevant level of order is now Y. This is illustrated in Figure 5.1.

The result is an excess of disorder H (equal to X – A) and some gain in order G (equal to Y – B). It is the work that produces the gain but entropy always increases. That is A < X, always. This is Rule C.

5.4 SOLAR ENERGY

Sunlight is a form of electromagnetic radiation (just like radio transmissions). It comprises not only the wavelengths of the visible spectrum but infra-red

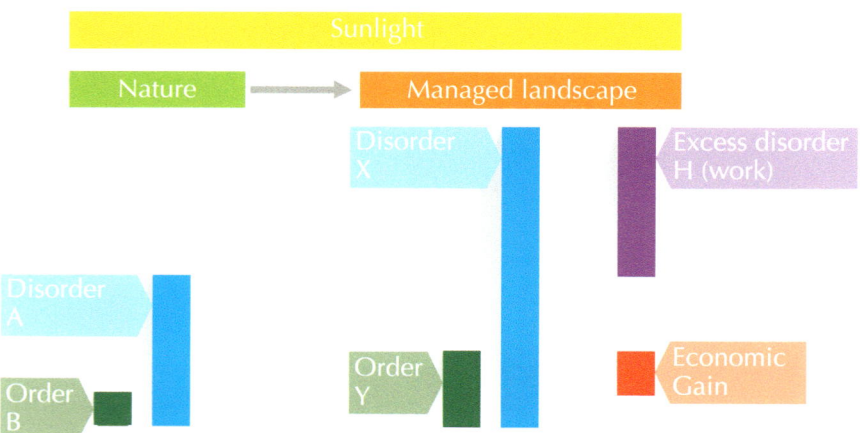

Figure 5.1 Entropy: order and disorder.
Credit: Nethergill Associates. Illustration: Elaine Leggett.

and ultraviolet radiation (Figure 5.2). Some solar radiation is reflected in the atmosphere by clouds and vapour and some more absorbed by the same. The degree of reflection is measured by the Albedo factor and this is calculated to be 30% (Figure 5.3). Aside from reflection, it passes through without

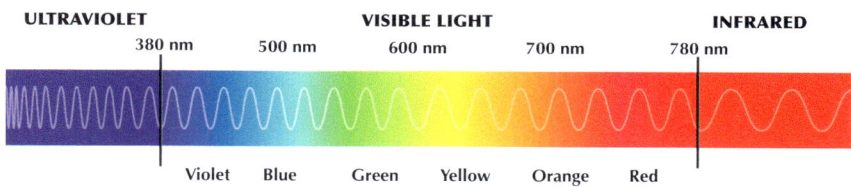

Figure 5.2 Light spectrum.
Credit: AdobeStock_458252033.

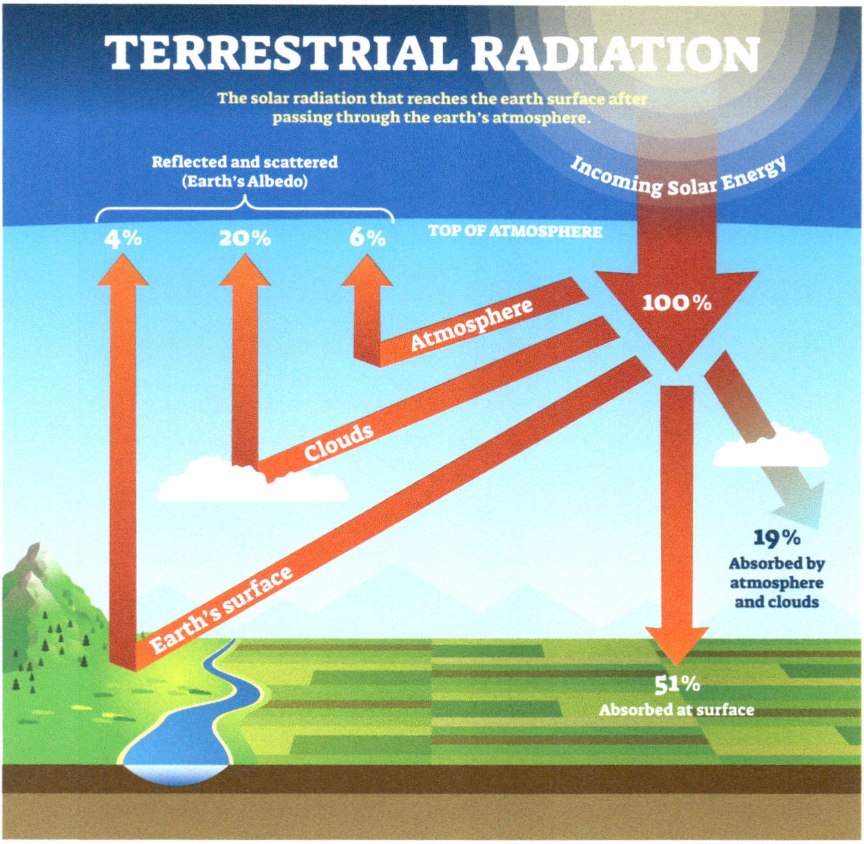

Figure 5.3 Terrestrial radiation.
Credit: AdobeStock_300026092.

exchanging much of its energy. However, when it strikes the ground, or indeed the human face, it is absorbed and re-radiated at a different wavelength – the infra-red wavelengths associated with heat. The direct heat from the sun warms the Earth to about –20°C and the re-radiated sunlight then warms the atmosphere up to its universal ambient level of 15°C.

Sunlight also drives the photosynthesis mechanisms in plant life. Photosynthesis produces O_2 (oxygen), which is essential for human and animal life (Figure 5.4). Our respiratory product is, conveniently, CO_2 and this is recycled by plant life. About two-thirds of the CO_2 consumed by plant life comes from recycled sources. Our very existence is governed by this virtuous circle.

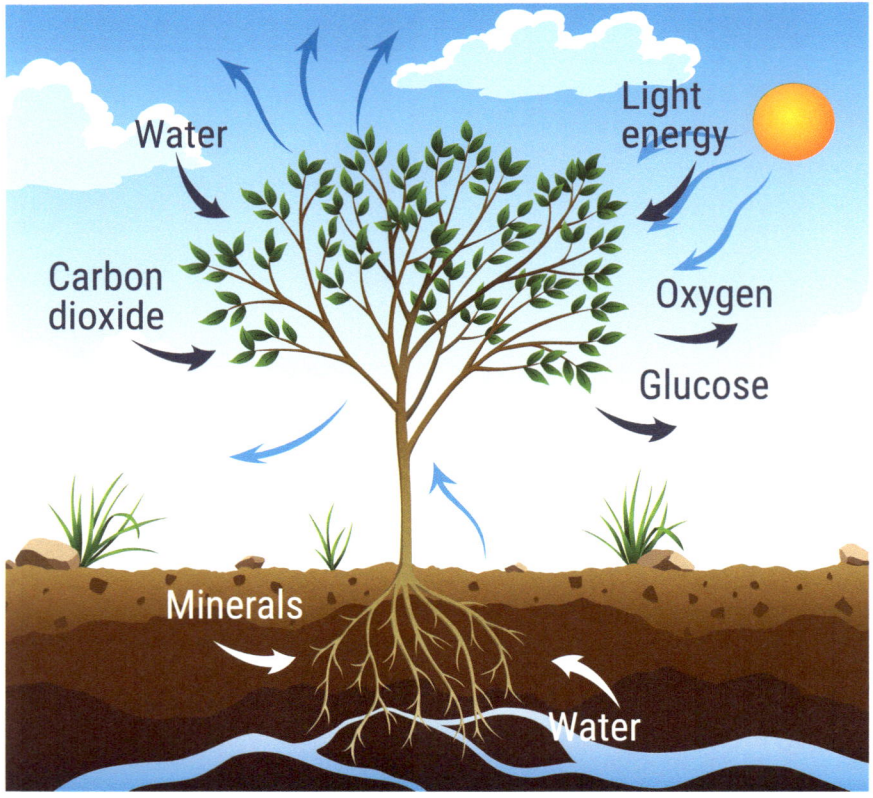

Figure 5.4 Photosynthesis.
Credit: AdobeStock_519075251.

The sun delivers, in the UK, an average of 101.2 W to *every* square metre.[4] It is this energy that drives all farming activities. In fact, it could be said that the primary objective of farming is to turn the incident solar energy available on a farm property into valuable food and other beneficial produce.

It is important to differentiate between the effects of the visible spectrum on photosynthesis and the heating effect of the sun's radiation. The importance of photosynthesis is in the role it plays in effecting chemical change within plants. Photosynthesis takes place whenever elements of the visible spectrum reach the surface and is affected by local cloud cover reflections. However, the rate of crop growth will be determined by the ambient temperature, which will be affected by considerations such as latitude.

5.5 INDUSTRIAL ENERGY

Despite having access to incident radiation, for photosynthesis, some farm properties will be naturally more productive than others. For example, the direct heating effect of the sun is influenced by latitude; some regions of the planet are excessively dry and others excessively wet as a consequence of climate differences; high elevation properties have different capabilities than low-elevation ones (sometimes better, but mostly worse).

Additionally, there has been a continuous quest to draw greater productivity from our natural resources in the form of better yields and lower unit costs. To correct for different economic disadvantages and to improve yields the industrialised world has developed a range of artificial substitutes such as fertilisers, herbicides and pesticides. These substitutes have certainly improved yields, but their true costs have not been recognised.

Firstly, all artificial substitutes have to be manufactured and manufacturing processes consume industrial energy. The industrial energy content in these substitutes can never be fully recovered in the food produced (as a consequence of the 2nd Law). The use of artificial substitutes has been justified on a marginal value theory – the price of food produce being sufficient to cover the cost of the industrial energy content. This, as will be seen, can be a flawed analysis. However, in times of crisis maximising food production may well be in the national interest, but this expedient will result in the decapitalisation of the managed landscape.

Figure 5.5 German chemist Fritz Haber, who developed a highly efficient ammonia process that was industrially scaled by Carl Bosch.
Credit: World History Archive / Alamy Stock Photo.

Nitrogen is notoriously inactive as an element. Although abundant in the atmosphere (N_2), accounting for 78%, and essential for all life, it remains difficult to 'fix' in useful products, such as ammonia (NH_3), which is the driving force behind the eventual formation of nitrates biologically and fertilisers industrially. The Haber–Bosch process (Figure 5.5), dating from 1910, has been transformative in the production of artificial fertilisers and is credited with huge advances in agricultural outputs worldwide. However, it also comes at great cost. The equation looks simple enough:

$$N_2 + 3H_2 = 2NH_3$$

The process takes nitrogen and hydrogen gases and forces them together under 200 atmospheres of pressure at temperatures in excess of 300°C over a bed of iron, which acts as a catalyst. The energy consumption is prodigious and this process alone is thought to account for 6% of world electricity production and 2% of the energy consumption of the world economy.[5]

Secondly, artificial substitutes have led to the development of a monocultural environment in those farming communities which have adopted a practice of heavy usage. Some of these communities now project a future limited to just 80 further harvests before sterility.[6] While this is widely contested the continuous degradation of soil quality, offset by the use of fertilisers at ever increasing rates, will at some point result in the inability to produce food physically and commercially. Sustainable farming must deliver perpetual fertility by contrast.

5.6 MSO AND ENERGY BALANCE

The MSO model is driven by underlying issues regarding energy. As free-issue energy is used, the 3rd contribution for a farm business climbs in proportion. When the free-issue energy is all consumed the farm will be at MSO and 3rd contributions will be maximised. As inputs with an industrial energy content are now consumed to push output, the 3rd contribution is reduced. The energy balances that produce this effect are shown in Figure 5.6

Free-issue energy is shown by the red line in Figure 5.6 (top). The pattern of 3rd contribution is shown by the green line and the industrial energy content of purchased supplies is the red dotted-line. It shows that the 3rd contribution peaks when all the free-issue energy is used. This is the MSO point.

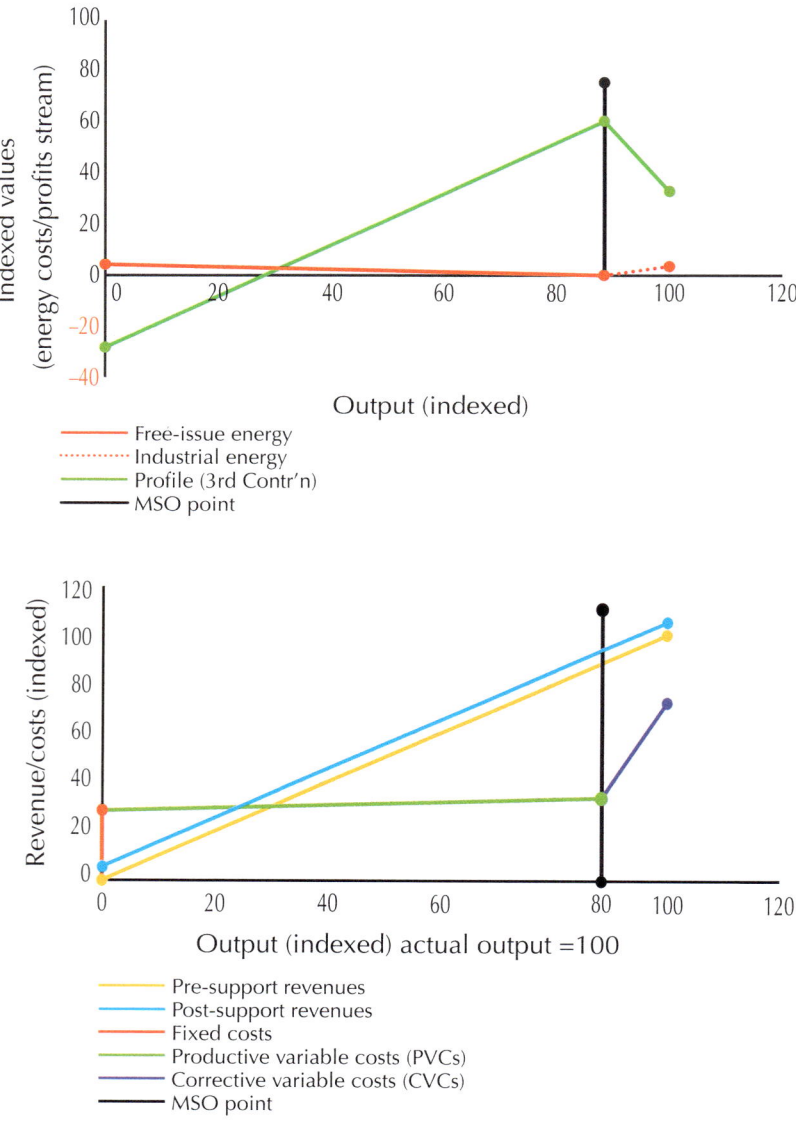

Figure 5.6 Energy balances and the MSO model.

Figure 5.6 (bottom) shows the corresponding break-even chart for the business on the MSO model and the exact alignment between the two diagrams. It is the realities of energy that drive the shape of the MSO phenomenon.

5.7 ENERGY BALANCE AND NET ZERO

Maximum sustainable productivity on a farm is delivered when a farm uses all its available energy from the sun, the free-issue energy resources, to produce food without any recourse to supplies that contain an industrial energy content. At this point, a farm will be, simultaneously, at its point of MSO; its point of maximum profitability (as its costs must be minimised by its being in its most productive configuration); and its point where biodiversity is not being compromised by environmentally polluting practices.

This point can also be argued to be its true net zero position as such a farm would have no industrial energy content. Intensive farming practices are entirely predicated on using industrial energy embedded in purchases and supplies to substitute for Nature. Farms using industrial energy content can never reach a true net zero position.

5.8 FARMING AS AN ENERGY PRODUCER

Let us define farming as an economic endeavour that maximises the use of the sun's energy (a free-issue, low-entropy energy resource) to produce grasses and grains for conversion into food products. Efficient farming will then, without any industrial inputs or interventions, use appropriate ruminants to convert grasses into meat products and use appropriate plant varieties to grow grains.

Livestock farming will be most effective when native breeds are adapted to an all-grass diet and year-round grazing outside. Nothing will be more productive nor more profitable.

Arable farming will be most effective when viable natural grains, being those adapted to the prevailing soil characteristics, climatic conditions and geological realities are farmed on a rotational basis, to maintain soil fertility, with help from the natural fertilisers the livestock produce. Here again, nothing will be more productive nor more profitable.

> **Responsible farming** will minimise the returns of slurry, phosphates, surplus nitrates, etc. (all high-entropy waste-products) to Nature.
>
> **Ethical farming** will deliver, additionally, uncompromised standards of animal welfare, and a perpetual horizon for soil fertility.
>
> Responsible and ethical farming at the MSO point will:
>
> - maximise farm profitability
> - maximise sustainable outputs
> - minimise the consumption of energy in the delivery of produce
> - eliminate pollution and environmental damage
> - produce a managed landscape in dynamic equilibrium with Nature where biodiversity is uncompromised and optimised for food production.

Farming nationally on this model will evolve into a very different pattern of land use. It has been estimated that as much as 40% of arable farming, in 2024, is dedicated to the production of animal food.[7] As MSO practices become more widespread this land will be available as more pasture for livestock or used for other arable products with the prospect of substantial growth in the value of national farming output.

CHAPTER 6
NATURAL CAPITAL

6.1 TYPES OF CAPITAL

Capital is a common concept that turns out to be remarkably complex and subtle. It can be classified into three family groups.

1 **Tangible capital** comprising natural capital, the primary concern of this book, and physical capital, of the sort associated with infrastructure.
2 **Intangible capital** comprising intellectual (or human) capital and social (or institutional capital).
3 **Artificial capital** that is financial capital created by convention for some social or economic purpose and comprises, usually, some forms of equity (or investment) and debt (or credit).

Only tangible capital, which is real in every sense, is constrained absolutely by the laws of physics and has to respect the ever-present phenomenon of decay. The other forms are unconstrained and would seemingly be free to grow indefinitely, without decay. However, as frequently demonstrated in the world of finance, artificial systems are prone to collapse when a sense of balance is lost. The critical distinction between tangible capital and the other two forms is that being constrained by the laws of physics, there is a **negative feedback** mechanism in place that ensures that nothing can be won 'for free'. The **positive feedback** mechanisms that characterise intangible capital and artificial capital, which act oppositely, will need close management if continuity is to be maintained and 'run-away collapse' situations are to be avoided.

> With respect to feedback mechanisms, negative does not mean 'bad' and positive 'good'. Negative (constraining) feedback mechanisms reduce the factor(s) they act on in a system. Positive (amplifying) feedback mechanism increase the factor(s) they act on in a system.

A common way to classify capital has emerged in recent years and comprises:

- natural assets
  - renewables
  - non-renewables
- physical assets
  - economic systems
  - monopolistic systems
- human assets
  - knowledge
- social assets
  - history
  - culture
  - values.

6.2 WHAT IS CAPITAL?

Essentially, capital is a store of value; a repository that can be called on from time to time to address situations that cannot be tackled by real-time actions. The concept of value is critical and makes it imperative that all capital must be capable of being measured and measured on a common basis in all its forms so that comparisons can be made.

The measure for all economic activity has evolved into a series of complementary monetary value systems and as natural capital has an economic value, in its social and environmental role, it, too, must be measured in monetary terms. This offends many of those parties that are sensitive to environmental degradation and sceptical of the perceived benefits of the commercial and financial worlds. Such groups are more comfortable expressing views about natural capital in purely qualitative terms. Unfortunately, such an approach destroys any basis to compare options rationally.

6.3 NATURAL CAPITAL

All our natural systems are driven by solar energy. Natural systems come in two basic forms:

- those driven by random processes
- those driven by chaotic processes.

Nature typically evolves by a series of random steps where choices are made based on the specific conditions that prevail at a point in time. When there is a negative feedback mechanism present, acting as a constraint, the outcomes of change can be estimated on the basis of probabilities but as the probabilities tend to be low for any particular situation, natural change is usually a complete surprise. This type of random change, often referred to as a **random walk**, will be irreversible (a consequence of Rule C) but will eventually reach a point of equilibrium (a consequence of Rule A). Equilibrium happens when the entropy of the system is maximised and, in practical terms, this equates to an abundance of biodiversity (which is a state of disorder).

When there is a positive feedback mechanism, which can happen locally, the randomness becomes chaotic. This happens when there is a quasi-regularity (technically, an *aperiodic* element) to a pattern of behaviour. Aperiodic systems are such that no previous state is ever repeated exactly. All natural systems when left alone will decay in the sense that order changes to increasing levels of disorder. The distinctive feature of chaotic behaviour is that at some point it can produce a phenomenon known as **resonance** where order is created out of disorder for some small part of the system. This is the process behind our geology and one of the best examples of the resonance phenomenon is the Giant's Causeway in Northern Ireland (Figure 6.1). Cooling lava, as with any cooling liquid, will self-organise at some point into hexagonal columns. The condition is transitory but if it coincides with some force that causes crystallisation evidence of the resonance will be left. Although the probabilities of such a coincidence are vanishingly small, on geological timescales the results can be seen everywhere.

In summary, then, natural capital is essentially our biodiversity (from random walks) and our geological resources (from chaos).

6.4 THE MEASUREMENT OF CAPITAL

Consider the case of a traditional type of quoted public company. Its capital value can be evaluated in one of three main ways.

Firstly, it will have issued shares and these shares will have market price. The *market capitalisation* of this company will be total value of the shares issued at the market price. This form of valuation is often used in take-over situations where a buyer will offer a premium over market value to attract sellers.

Secondly, it will have a *net assets value* on its balance sheet. This will be the net valuation of physical assets value, debt liabilities and intangible asset

Figure 6.1 The Giant's Causeway.
Credit: AdobeStock_549424399.

values related to issues such as reputation and intellectual property. While this might seem to be the most obvious way to value a company it is a surprisingly poor guide except in cases of bankruptcy.

Thirdly, it will have a value to its future income stream from profits. This method reflects a process which addresses the question:

What should I pay, in one single payment up-front, for the rights to a stream of income (the profits) over a number of years (essentially in perpetuity for an on-going business) bearing in mind that a future income stream cannot be predicted with precision?

This is the net present value of the company. It is the most effective measure of real value for a company and this should be the model applied to the valuation of natural capital.

6.5 THE CONCEPT OF NET PRESENT VALUE

Suppose a farm business projects its most likely *minimum future profits* over the next 5 years to be £20,000, £25,000, £30,000, £25,000 and £25,000. Notice that the projections fall in value after the third year. This is normal when making projections as uncertainties grow in later years. In reality, there might be an expectation that profitability would continue to grow but this, of course, cannot be guaranteed.

If these projections are added up, making a total of £125,000, this could be regarded as the value of the business over the next 5 years. A seller might ask this as a price for the business if it was for sale, but a buyer might argue that such a price would imply that he or she would be paying today for 4 years of future income. Not only is that income projection uncertain, but the buyer does not have its full benefit today. If interest rates are 5%, he or she could argue that the projection for the second year, £25,000, is only worth 95% of the projected value today. Similarly, the projection for the third year, £30,000, is only worth 95% of 95% (90.25%) of the projected value, and so on. This process is known as **discounting**.

If the profit projections are now discounted for each of the 4 years in the future, the sum of these discounted values is called the **net present value** (NPV) of the profits stream over the five-year horizon. At 5% interest rates for the 5 years, the NPV of the profits stream is £113,184 as compared to the simple total of £125,000. The NPV would represent a fair price but the buyer would be gambling on the projections being realised and hoping that the profit stream in the later years would be much improved as today's uncertainties become resolved positively.

These calculations are not difficult but can be quite tedious. If we take the NPV, of £113,184, and divide it by the first year's profit projection, £20,000, we get 5.66. This number is known as the PE ratio (or price to earnings ratio) and is a shorthand form of evaluation. If this number is small, say 2, the quality of earnings is said to be low. If the number is large, say 12, the quality of earnings will be deemed to be high.

The PE ratio is a reflection of the quality in the income stream of a business. High PE ratios are associated with companies that have long-term contracted or predictable income streams, such as utility companies (e.g. with a 25-year supply contract for electricity), or highly speculative growth companies (e.g. a software technology company). Low PE ratios are associated with companies that experience great uncertainties in their future income streams as a result of factors such as weather, politics or short horizon contracts (e.g. a building contractor where contracts may last 18 to 24 months with no guarantee of any future work). In the first case a PE ratio in excess of 15 may be found and in the second case the PE would more likely be 1.5 to 2.

PE ratios are a judgement call and over time a consensus often emerges for a business sector. In this book, we shall adopt a PE ratio for farm businesses, which will have some ability to project future earnings, but low expectations of profitability, of 4.

6.6 PRIMARY NATURAL CAPITAL

To value natural capital, it is first necessary to define what income can be attributed to the ownership of natural assets. In other words, what is Nature's contribution to earnings on a farm? The primary income (PI) stream on a farm is taken to be:

PI = pre-support revenues – productive variable costs (PVCs)

This is the revenue from the sales of produce less those inevitable costs in working with Nature so as to produce. The NPV of this income stream is valued at 4 × PI. This is the primary natural capital (PNC) value of a farm business. Therefore:

PNC = 4 × (pre-support revenues – PVCs)

Note that this is not the value of the profits stream. To evaluate the 1st contribution, the CVCs have to be deducted. When this is done the residual natural capital value is reduced and this is termed the secondary natural capital of the business; to value the 2nd contribution there has to be a further reduction to account for fixed costs, and the remaining natural capital value is reduced further and is termed the tertiary natural capital of the business. These reductions are called attenuations. As costs are incurred after the PVCs the effect is to decapitalise the natural capital value of the farm business. The natural capital evaluations for a small upland farm is shown in Figure 6.2.

It can be seen from the evaluations that natural capital is destroyed very quickly as CVCs and fixed costs are incurred. For the farm illustrated in Figure 6.2 tertiary natural capital value is negative implying that the farm is drawing on its natural resources and depleting them in commercial terms. This situation is being masked by the subsequent addition of support payments, although if these are deemed by society to be a form of price support the negative value, in this case, would turn positive.

6.7 INTRINSIC NATURAL CAPITAL

There is a popular view that there is some *intrinsic* form of natural capital; a value relating to the animal, vegetable and mineral resources of Nature

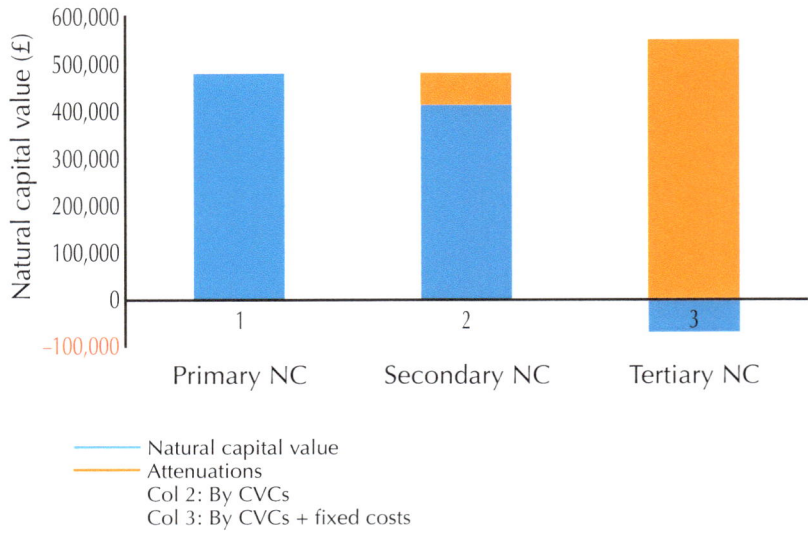

Figure 6.2 Hierarchy of natural capital evaluations.

and its landscape, which should be expressed in a non-monetary form suggesting it is more valuable than mere monetary considerations. This approach can only be qualitative and suffers from the inability to make valid comparisons between different situations. This is exemplified by the problem in answering the question: how many worms equal a lapwing? There will never be any real agreement on such relationships when more than two people are involved or when more than two natural features are present. Intrinsic natural capital of this type can only reflect personal or subjective views.

6.8 ACCOUNTING FOR NATURAL CAPITAL

If measures of natural capital are to play a full part in business activities they will have to be incorporated into the management accounts of the business. Measures of capital are properly accounted for on the balance sheet of a business. Interestingly, even in the largest of corporations it is unusual to have a balance sheet cast for management accounting purposes (as opposed to statutory reporting). However, a management accounts version is indispensable in directing the use and deployment of business assets in the most advantageous or most cost effective way.

The traditional source of capital in a business, see Chapter 2, will be investors and subscribers. The funds provided by these sources are classed as business liabilities. To accommodate concepts of natural capital the first step is to recognise Nature as a stakeholder in a farm business. As such, its contribution must rank as a liability alongside other stakeholders. The other, traditional, stakeholders will have put in funds and will expect a dividend in due course for the use of those funds. In contrast, Nature gifts its resources to the property owner and, rather than imposing a burden on the business for a dividend, it actually provides a free-issue bounty. As traditional stakeholders appear on the balance sheet as *positive* liabilities, Nature must be a *negative* liability. At first this will all seem counterintuitive.

Whatever natural capital value is ascribed to the business it will have an equal and opposite entry on the assets side as natural assets employed. This will be negative, too, and it will have the effect of reducing the net assets value employed in the business. That is, as the natural capital of a business increases it improves ROTA (as total assets employed reduce in net value). This is as it should be. Now when natural capital values in a business fall, there will be decapitalisation and performances will suffer. Also, in growing the natural capital of a business the easier it becomes to strike a particular ROTA target. In other words, managing Nature responsibly will be a profitable endeavour.

6.9 CASE STUDIES: NATURAL CAPITAL CONSIDERATIONS

Consider the businesses described by the B/S set out in Figure 2.1, Farm A and the P&L set out in Figure 2.2, Farm B. Their key data is summarised in Figure 6.3 together with a performance analysis to demonstrate the impact of natural capital considerations.

As Farm A produces a positive 2nd contribution, its ultimate (tertiary) natural capital value is the consequence of its status as a negative liability. The

Figure 6.3 Impact of natural capital on performance.

Farm business	A	B
Pre-support sales	2,240,000	220,000
2nd contribution	336,000	-10,000
Support payments	560,000	55,000
3rd contribution	896,000	45,000
Turnover	2,800,000	275,000
Assets employed (conventional)	5,600,000	850,000
Tertiary natural capital	−1,344,000	40,000
Effective assets employed	4,256,000	890,000
Performance	**A**	**B**
CONVENTIONAL		
Margin 3rd contribution (%)	32.00	16.36
Assets turn (n)	0.50	0.32
ROTA	16.00	5.29
WITH NATURAL CAPITAL		
Margin 3rd contribution (%)	32.00	16.36
Assets turn (n)	0.66	0.31
ROTA	21.05	5.06

effect is to deliver an improved ROTA. This is the benefit of growing natural capital.

As Farm B produces a negative 2nd contribution its ultimate natural capital value serves to put an extra burden on the business to produce a return and ROTA is reduced. This is the penalty for decapitalising its natural assets.

Moving to MSO will always improve the 2nd contribution. This is inevitable as CVCs will have been eliminated. This improvement can transform the situation regarding natural capital. The transformation for a medium-sized arable farm is set out in Figure 6.4.

Initially, as output is reduced in moving to MSO, the PNC of the farm decreases. However, at MSO the secondary natural capital is un-attenuated as all CVCs have been removed. When this is reflected in the ultimate (tertiary) natural capital it has the effect of reversing the decapitalisation.

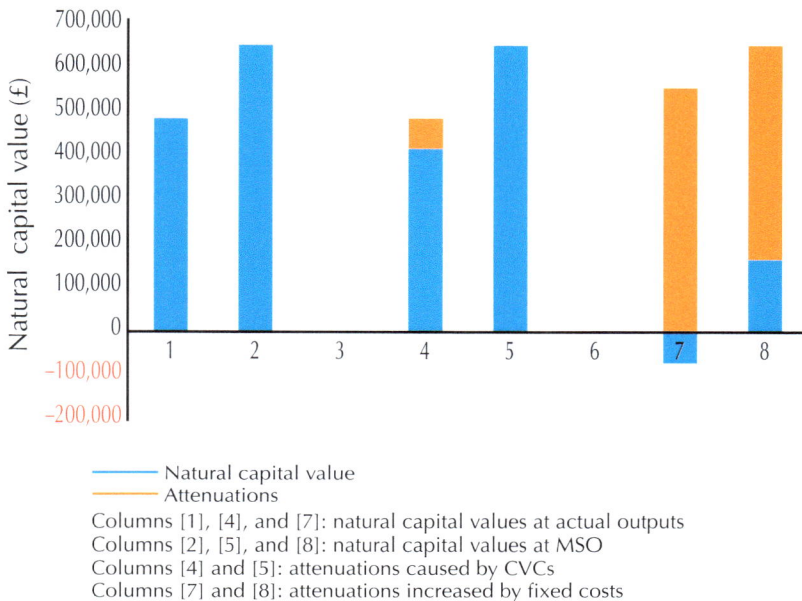

Natural capital value
Attenuations

Columns [1], [4], and [7]: natural capital values at actual outputs
Columns [2], [5], and [8]: natural capital values at MSO
Columns [4] and [5]: attenuations caused by CVCs
Columns [7] and [8]: attenuations increased by fixed costs

Figure 6.4 The impact of MSO on natural capital values.

This accounting convention is not yet acknowledged for statutory accounts but that does not inhibit its use in management accounts which should be focused exclusively exercising on management control.

CHAPTER 7
MSO THEORY

7.1 SEQUENTIAL VARIABLE COSTS

To explain the phenomenon that a number of farms were clearly more profitable when the scope and scale of their business activities were reduced
it was necessary to create a variable cost line that was bifurcated (that is
in two sequential parts). Only then could a business model be created
that would offer the prospect of breaking-even, at some volume, and then
threaten to break-back, at some later volume. This configuration of costs is
not an approximation of some single non-linear cost function; the sequential
nature is its critical difference.

Standard economic theory covers cases of non-linear cost functions quite
adequately with the help of differential calculus. This success is behind the
sentiment, often expressed, that the law of diminishing returns is one of the
great successes of economic theory. The bifurcation models do not challenge this; the models simply introduce a new pattern of behaviour alongside, but not instead of, the concept of diminishing returns.

Some economists have argued with the authors that even if variable costs
are de-composable into productive variable costs (PVCs), based on working
with free-issue natural energy sources, and corrective variable costs (CVCs),
based on working with natural substitutes with embedded industrial energy
components, farmers will not differentiate in their usage in practice. As a
business expands it will be unable to differentiate between PVCs and CVCs
and proceeds along a composite total variable cost line. However, when

it comes to downsizing by the elimination of CVCs it will follow the CVC line to its MSO point. This process is shown in Figure 7.1 where expansion follows the red dotted arrow but downsizing will follow the orange dotted arrow.

7.2 THE IMPORTANCE OF ENERGY BALANCES

The decomposed variable cost function is a reflection of the underlying impact of dealing with two very different sources of energy. Whatever patterns of activity emerge on a farm there has to be an energy balance; the consumption of energy by livestock or by cereals has to be met by a corresponding supply of energy. When consumption exceeds the ability of Nature to supply, the deficit must be met by drawing on other sources, which will inevitably involve artificial supplies produced with industrial energy.

In the case of livestock farms, the principal challenge is to maximise the production of good quality grass and to have it available throughout the year for feeding. In the event that there is insufficient grass available for the

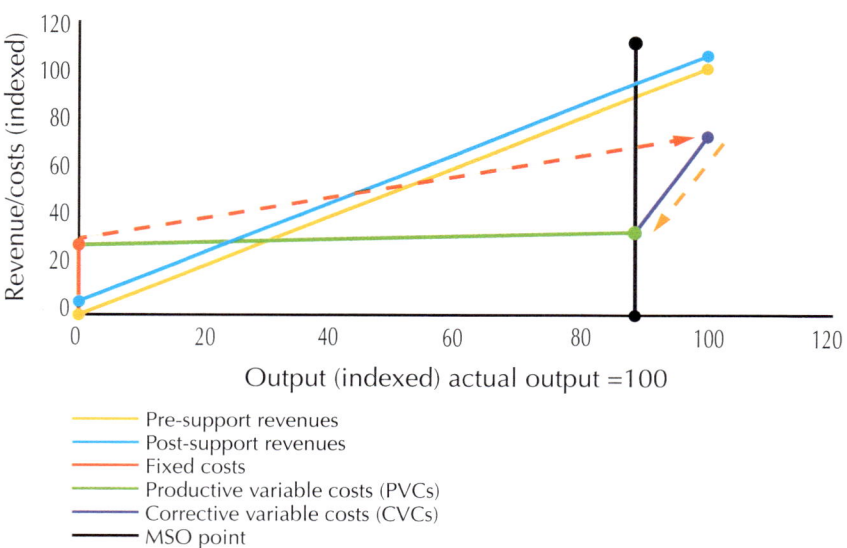

Figure 7.1 Practical realities of PVCs and CVCs.

livestock on a farm, a farmer may, either buy-in feed concentrates to cover the deficiencies or use fertilisers to produce a greater volume.

Consider a livestock farm business in the north of England that produced grass with an energy content of 20 million MJ. This cost £8,500 to produce and harvest = 2,352.9 MJ/£. It proved to be insufficient on its own to serve the herd and the flock and it was felt necessary to purchase additional feed concentrates rather than reduce stock numbers to compensate. The farm purchased concentrates with an energy content of 235,000 MJ at a cost of £25,000 = 9.4 MJ/£. The asymmetric nature of these two different expenses determined the ultimate profitability of the farm business.

In the case of arable farms, the principal challenge is to satisfy the demand for nitrogen in growing crops (see Section 7.4 for a specific example). Balancing the demand for and the supply of nitrogen is producing an energy balance in another form. Nitrogen comes from four principal natural sources: from mineralisation processes in Nature, from the production of nitrates from the natural chemistry of soils, from the decay processes of dead matter, and from manures. A very small amount will come from the impact of lightning in the atmosphere. Artificial sources of nitrogen in the form of fertilisers mostly come in the form of ammonia (NH_3) which, in turn, mostly comes from a version of the Haber–Bosch process.

The Haber–Bosch process has been cited as one of the great advances underpinning economic growth worldwide. It has certainly contributed to increased farm outputs by increasing the yields in arable farming. In the hundred years from 1923, wheat yields in the UK improved from about 2.4 tonnes per acre to about 8.2 tonnes per hectare. However, this improvement has been hard won. It takes, at best, 15,000 MJ of energy to produce a tonne of ammonia by the Haber–Bosch process. This is equivalent to 2% of the energy consumption of the world economy and it is responsible for an estimated 1.8% of all carbon emissions.

The MSO point on a farm will always coincide with an energy balance that has no industrial energy component.

7.3 MSO FOR LIVESTOCK FARMING

Every breed of animal has a typical dietary requirement. For beef cattle this is about 35,000 MJ per head per annum; for lambs it is about 2,000 MJ per head per annum. Using the relevant rates for the stock on a farm it is simple to establish a total dietary requirement. This requirement will be met, principally, from grass which will have a typical energy density of 11.0 MJ/kg dry matter. In this way, the grass requirement to sustain the livestock on a farm can be established.

An example of this calculation for a mixed livestock farm is shown in Figure 7.2. The livestock on the farm, rated at 483 livestock units (LSUs) has a requirement for 1,011,364 tonnes of dry matter.

Every different type of grassland produces different amounts of forage. If a standard value of 7,500 kg dry matter grass per annum is taken to represent a rotational grazing regime then other types of grassland can be rated against this. Permanent pasture could then be rated at 70% and rough grazing (grade 3) could be rated at 20%. Using the relevant rates for the different areas of grassland that comprise a farm it is again simple to calculate the production of grass for grazing.

This calculation for the livestock farm, comprising 483 LSUs, is shown in Figure 7.3. It will be seen that this farm, of 170 ha, is capable of producing 885,000 kg of grass.

> The MSO ratio on this farm is the ratio of grass produced to grass requirement, in this case 0.88 (= 885,000/1,011,364). That is, the MSO point on the farm is at 88% of current output levels and this would be achievable if all its CVCs were eliminated.

Figure 7.2 Model calculations: grass requirement for a livestock farm.

Livestock				Grass consumption		
	Number	LSU rating	Effective	MJ/hd pa	MJ/kg DM	kg DM pa
THE HERD						
Beef cattle	80	1.65	132.00	35,000	11.00	254,545
2 year olds	85	1.20	102.00	25,000	11.00	193,182
Calves	85	0.65	55.25	16,000	11.00	123,636
Dairy herd (milking cows)		2.00	0.00	45,000	11.00	0
Calves (concentrate-fed)		0.40	0.00	18,000	11.00	0
Effective herd			289.25		Total	571,364
THE FLOCK						
Ewes	600	0.20	120.00	5,000	11.00	272,727
Lambs	920	0.08	73.60	2,000	11.00	167,273
Effective flock			193.60		Total	440,000
Effective livestock			**482.85**		**Total**	**1,011,364**

Figure 7.3 Model calculations: grass production on a livestock farm.

Class	Rating	Area (ha)	Effective area (ha)
Rotational	1.00	30.00	30.00
Long term	0.90		0.00
Perm pasture	0.70	40.00	28.00
Forage	1.00		0.00
Rough grazing 1	0.60	100.00	60.00
Rough grazing 2	0.40		0.00
Rough grazing 3	0.20		0.00
Totals		170.00	118.00
Standard (kg DM/ha pa)			7,500
Total (kg DM pa)			885,000

7.4 MSO FOR ARABLE FARMING

The primary requirement in arable farming is to maximise the availability of natural sources of nitrogen (N_2). Most arable farms go beyond this point and add N from fertilisers. A typical fertiliser will contain 130 l-N per tonne and wheat will require 279 kg-N/ha to deliver a standard yield of 7.50 tonnes/ha. The N-balance on a farm can be established from these parameters when the outputs are measured, and the application rates of fertilisers are recorded.

An example of the calculation for a cereal farm is shown in Figure 7.4. This farm comprised 1,100 ha and produced winter wheat with an energy value of 19.3 MJ/kg. Total output was 8,250 tonnes and 1,200 tonnes of fertilisers were purchased.

> The calculation shows that fertilisers were delivering 57.75 kg-N/ha and that 221.25 kg-N/ha was supplied from natural resources. The MSO ratio on this farm is the ratio N produced from the soil to total N consumption, in this case 0.79 (=221.25/279.00). That is, the MSO point on this farm is at 79% of current output levels and this would be achievable if all its CVCs were eliminated.

Figure 7.4 Model calculation: the nitrogen balance on an arable farm.

Land	1,100	ha
Output	8,250	t
Fertiliser use	1,200	t
Application rate	0.92	t/ha
N-content [F]	130.00	l/t
N-delivery [F]	57.75	kg/ha/pa
N-requirement	279.00	kg/ha/pa
MSO-datum	221.25	kg/ha/pa
MSO ratio	0.79	

7.5 MSO FOR DAIRY FARMING

A dairy farm comprises two separate businesses – the livestock farm and the dairy parlour. The dairy parlour is a *factory*, in essence, and will follow the standard model of the firm in terms of economic behaviour. That is, it only has one category of variable costs which are always of the CVC-type for factories.

The livestock farm element of the dairy business will exhibit the MSO characteristics of all other livestock farms.

Milk output rates per cow can be driven by the use of supplementary feed concentrates and, currently, few dairy herds are all-grass fed. As output rates per cow are pushed up so the useful life of the cow as a milk producer declines. This useful life is measured in lactation cycles. The essential trade-off between output rates and useful life is shown in Figure 7.5 which is based on the anecdotal evidence of different dairy farmers.

The lifetime output of milk per cow would, on this basis, appear to be maximised at 6,000 l per annum and a useful life of 6 lactation cycles. However, this output is unlikely to be achieved with an all-grass feed and some feed concentrates will be most likely. The relationship between concentrates

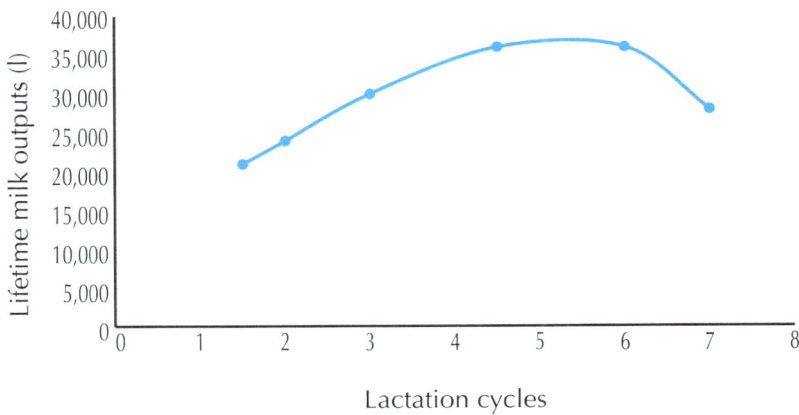

Figure 7.5 The trade-off in milk yields.

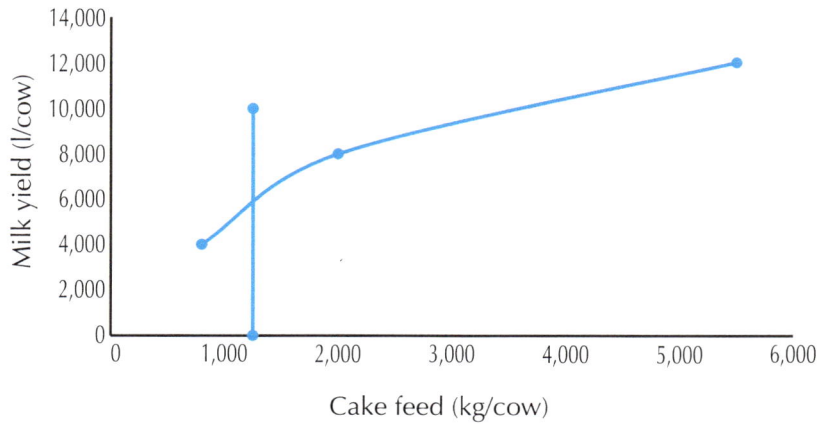

Figure 7.6 Milk outputs and concentrates usage.

usage and milk output is shown in Figure 7.6. At 6,000 l/year output the most likely rate of concentrates use is about 1.25 tonnes/cow/year.

On this basis, the target for optimum output in a milk parlour would be:

> Output rates (l/cow/year): 6,000
> Milking life (Lactation cycles): 6
> Concentrates usage (tonnes-cake/cow/year): 1.25

This prescription is optimal in the sense that it will maximise the output of a cow over its economic lifetime and, simultaneously, minimise the unit cost of milk production and maximise the 2nd contribution per animal. This will not be the position of maximised revenues (see Chapter 8).

7.6 PIGS AND POULTRY

Pigs and poultry are all farmed with its entire feed supplies being purchased. These farms consequently follow the standard economic model of the firm and its factory model, where the single category of variable costs are of the CVC type.

However, the supply of grains will come from farms which are not subjected to the MSO model but it is unlikely that the pig and poultry businesses would be aware of its suppliers' behaviour in this regard.

7.7 MSO: THE CRITICAL ROLE OF INDUSTRY STANDARDS

The MSO calculations require access to critical industry sector standards to be evaluated with precision. The relevant standards used in these calculations are:

- for livestock farms
  - dietary requirements for animals by age and by breed (MJ/head/year)
  - energy content of grasses by type (MJ/kg)
  - standard grass production rate by grassland type (t/ha)

- for arable farms
  - the N-content of fertilisers by type (N-kg/t)
  - the N-requirement for cereal growth by type (N-kg/ha)
  - the energy value of cereals by type (MJ/kg).

These standard values are not generally available. However, a long established common source of reference is the *John Nix Pocket Book for Farm Management*. None will have the underlying rigour that would be normal for industrial engineering standards. It is an issue for a prospective sector-wide initiative. At a farm level, local parameters can be established with some work; the authors have relied, additionally, on a proprietary algorithmic model based on financial/operating data.

For the dairy parlour, more evidence would be helpful to refine the relationships shown in Figures 7.5 and 7.6 to include differences in breeds and differences in branded concentrates.

The application of standards should always be one of continuous refinement. As more and better data becomes available it should be reflected;

however, in the meantime using the best available, with qualifications, is better than nothing.

7.8 THE EXTRAORDINARY PROPERTIES OF THE MSO POINT

When a farm is at its MSO point it is possible to claim that:

- the 2nd contribution will be maximised
- the pattern of biodiversity in the managed landscape of the farm will be uncompromised and optimised (locally) for production
- the energy footprint of the farm, by eliminating all inputs with an industrial energy component, will be minimised
- the natural capital value of the business will be maximised.

These are four very powerful claims. When at, or very near, MSO the key strategic issue is to position the business so as to grow the MSO point and exploit, to the greatest extent, the improvements to be expected in soil fertility up to its natural limits.

Evidence from farms working on MSO implementation programmes already indicate that the MSO point grows as soils recover from fertiliser use or over-grazing and as breeds re-adapt to all-grass diets. One livestock farm in Wales cut out all its CVCs at a stroke, prompted by the COVID pandemic and its effects, and saw its grass production fall to 45% of its previous level. In the subsequent 4 years the output has climbed back to better than 95% of its former level but without the additional costs. An arable farm in eastern England is cutting out its CVCs progressively as part of an MSO programme and is already reporting a measurable increase in the depth of soil.

7.9 MSO AND FOOD SECURITY OF SUPPLY

It seems probable that, based on the evidence of the Nethergill Database, that currently, 93% of farms in England and Wales are working above MSO.

The average MSO point for these farms would seem to be about 85% of their current outputs. What are the implications on national food supplies, therefore, in a general move to MSO by the sector?

While outputs would immediately drop, there is every prospect that these would recover to near former levels. However, in addition, two other opportunities arise.

Those farms below MSO might be encouraged to expand production. Many of these farms are consciously below MSO for environmental reasons that appeal to their owners and they may be persuaded to expand. However, it is worth noting that on all the farms operating below MSO examined by the authors (over 20 at the time of writing) none have been free of CVCs!

Those farms that produce grains for animal feed will be able to move to other crops or convert to livestock. The Wildlife Trust estimates that 40% of arable production in England and Wales would fall into this category.[8] A move to MSO should result in an increase in food production as a consequence in the course of time.

CHAPTER 8
MSO IN PRACTICE

8.1 A HIERARCHY OF OBJECTIVES

A hierarchy of profit objectives were set out in Chapter 3 in the form of a succession of contribution levels. These are part of a wider set of business objectives.

> The first objective of a business is not to lose cash on every transaction.

This happens when revenues are insufficient to cover variable costs (where the 1st contribution would be negative). Ideally, a business should never lose cash in this way on *any* transaction. This test on a farm business needs to be made with revenues taken before support payments are taken into account. From the evidence of the authors' analyses up to 20% of farms in the UK might fail this first test.

> The second objective of a business is not to decapitalise.

This happens when revenues are insufficient to cover variable and fixed costs (where the 2nd contribution would be negative). Continuous deficits at this point in mainstream (non-farming) businesses will have to be covered by constant injections of capital funds to survive, either from new money or from the realisation of assets held on the B/S. Again, from the authors' evidence up to 80% of farms in the UK will fail at this point.

> The third objective of a business is to deliver a proper financial return on the assets employed (ROTA).

The immediate target for the 2nd contribution is to secure, at least, a 15% ROTA performance. This would beat a risk-free deposit of an equivalent amount to the assets employed in a bank, at say 5%, and provide a premium of 10% to compensate for the business risks.

> The fourth objective of a business is to produce a living wage for its partners.

The target for a suitable draw, out of the 2nd contribution and after delivering a 15% ROTA, is a matter of judgement for the partners but it does not seem unreasonable to expect, at least, a multiple of 2 to 3 of the statutory National Living Wage. Currently, in 2025, this would equate to between £50,000 and £75,000 per annum based on a 40 hour week.

> The fifth objective of a (farm) business is to minimise its dependency on support payments.

Support payments are a unique feature of farm businesses. These payments when added to the 2nd contribution will comprise the 3rd contribution. They are transformative in practical terms with 80% of farms in the UK being positive at this level. However, the BPS inherited from the Common Agricultural Policy (CAP) of the European Union is being phased out and replaced progressively by a new Environmental Land Management Scheme (ELMS). The essence of the change will be a move from an area payments system to one related to more direct management of the environment. At the time of writing full details have yet to emerge.

> The sixth objective of a business is to avoid being over-extended in servicing long-term debt.

This is easily done in farming as most businesses have, at least superficially, the ideal assets as collateral in their land. A good rule of thumb is to take 25% of the 2nd contribution and establish what capital advance might be secured if this sum was to cover the interest payments. Suppose the 2nd contribution was £100,000. A loan advanced on this basis at 5% interest would be £500,000 (£25,000 pa in interest payments). The relationship between the interest payments made and the level of the 2nd contribution is known as the cover ratio. Ideally, this should always be 4 or above. That is the interest charges are covered by the 2nd contribution 4-times over. One of the unintended traps in taking on bank loans is that some bank managers will advance money on their assessment of the free cash flow in the business after the 3rd contribution. This would make the ability to service the loans vulnerable to losses in support payments.

> The seventh objective of a (farm) business is to grow its natural capital valuation continuously.

Natural capital has been defined as the capitalised value of revenues less PVCs (see Chapter 6). This objective will guarantee that progress is being made for all the right reasons – commercial and environmental.

8.2 CASE STUDY: LIVESTOCK FARM, NORTH OF ENGLAND

> Consider the performance of a farm of 600 ha in the north of England with 350 ewes and 50 suckler cows. As the farmer owns the land, the business is split into two components: a landowning business, which charges an imputed rent to the farm operation; and a farm business, which operates on a tenanted basis.

Its MSO ratio was established at 0.85 and its management accounts (for the farm business on a tenanted basis), at actual and MSO, is set out in Figure 8.1.

By downsizing to MSO, through the elimination of its CVCs, the 2nd contribution margin improves from –24.64% to +3.26%. This would put the farm, just, in the top 20% of farm businesses in the UK. The improvement, while worthwhile, is not sufficient to produce a 15% ROTA nor provide enough surplus for a proper draw. However, fixed costs are equivalent to 73.21% of

Figure 8.1 Simplified management accounts for a mixed livestock farm in the north of England.

Management accounts (as a tenant)	Actual	At MSO
SALES		
Lamb	105,000	
Beef	35,000	
Total sales	**140,000**	**118,357**
COSTS		
PVCs	12,000	12,000
Income stream from Nature	128,000	106,357
CVCs		
Concentrates	26,000	
Fertilisers	11,000	
Others	23,000	
Total variable costs	**72,000**	**12,000**
1st contribution	68,000	106,357
Fixed costs	102,500	102,500
2nd contribution	–34,500	3,857
Support payments	120,000	120,000
3rd contribution	85,500	123,857
Fixed assets	300,000	
Stock	120,000	
Assets employed	420,000	420,000
Performance measures		
Margin: 2nd contribution	–24.64%	3.26%
Assets turn: 2nd contribution	0.33	0.28
ROTA: 2nd contribution	–8.21%	0.92%
Margin: 3rd contribution	32.88%	51.96%
Assets turn: 3rd contribution	0.62	0.57
ROTA 3rd contribution	20.36%	29.49%

its pre-support revenues and this is now the main problem. Evidence from the work of the authors suggest that it is very unlikely that a farm business will be positive at the 2nd contribution level if its fixed costs are in excess of 40% of its pre-support revenues. Reducing fixed costs is a very difficult challenge for a small farm and the future for this farm lies in the opportunity to grow by aggregation. That is, to grow by acquiring, when opportunities arise, neighbouring farmland. Initially, this expansion should be through renting or leasing, if possible, to minimise the assets employed in the business.

When the farm business is split into its two component strategic business units (SBUs) of sheep and cattle, the cattle is the more profitable part of the enterprise. This is shown in Figure 8.2. This is often, but not invariably, the case and farms with a cattle-to-sheep ratio of about 60 : 40 in LSU terms seem to do best. This is possibly the effect of maximising the take-up of available grass, especially when the sheep are grazed after the cattle.

When the MSO model is applied to the farm business it produces a distinctive pattern of revenues and costs. These are encapsulated in its break-even analysis. The farm business does not quite reach its break-even point in its configuration without CVCs. This is shown clearly in Figure 8.3.

The principal barrier to this being achieved is the level of fixed costs in the business. A modest reduction in these, say 5% (equivalent to a reduction of £10,000), would ensure a break-even point on pre-support revenues in the absence of CVCs. When all the free-issue resources have been consumed (at the end of the PVC line) there is an inflexion point in the behaviour of

Figure 8.2 Strategic business unit performance for a mixed livestock farm in the north of England.

SBU performance	Sheep SBU		Cattle SBU	
	£	%	£	%
Farm output	105,000	100.00	35,000	100.00
PVCs	10,200	9.71	1,800	5.14
CVCs	45,550	43.38	14,450	41.29
1st contribution	49,250	46.90	18,750	53.57
Fixed costs	76,875	73.21	25,625	73.21
2nd contribution	−27,625	−26.31	−6,875	19.64

its variable costs. This is the MSO point for the farm. Beyond this point only CVCs will be consumed. It can be seen from Figure 8.3 that CVCs rise faster than the growth in revenues. In this case, for every £1 of revenue won above the MSO it is costing the business £2.77. This is the adverse leverage of CVCs on profitability.

It is important to point out at this stage that farmers will tend to consume PVCs and CVCs in a composite fashion from the very start of activities; it is most unlikely any will consciously separate the PVCs from the CVCs. For this reason, some economists dismiss the decomposition of variable costs as a valid analysis, but it must also be recognised that the logic of the sequential nature of the cost components does not disappear and the reality of this is manifested in two ways. Firstly, at some point the PVC components will be exhausted in reality and, secondly, to move to MSO the practicalities are that this can only be done by a progressive reduction in CVCs. In other words, the separations are real when downsizing.

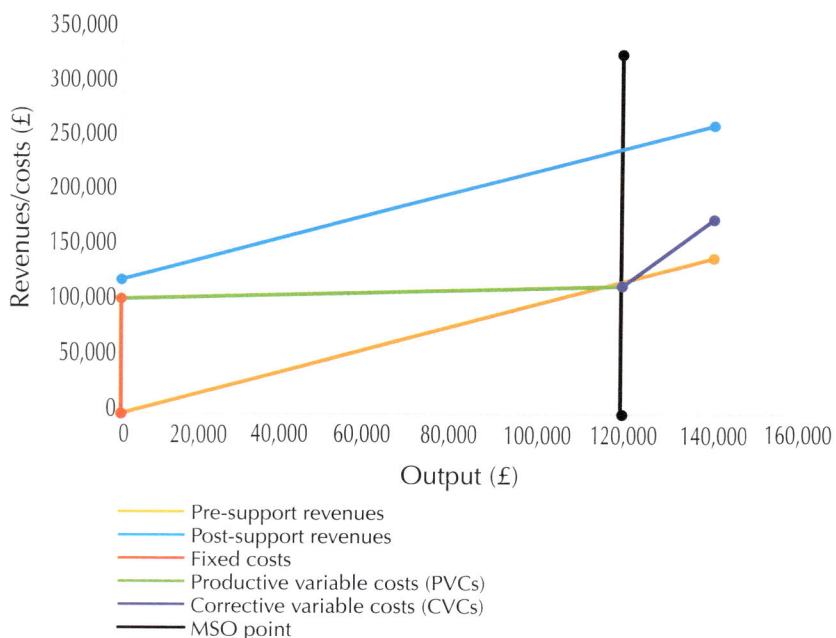

Figure 8.3 Break-even analysis for a mixed livestock farm in the north of England.

Driving the pattern of behaviour shown in the break-even analysis is an underlying pattern of energy use. This, too, is distinctive and is entirely responsible for the characteristics of the break-even performance. The underlying energy balance diagram for this farm is shown in Figure 8.4.

As the free-issue energy resources are used up, the farm business will grow in terms of increasing levels of profitability. When these free-issue resources are exhausted, energy from artificial substitutes (supplies with an industrial energy content) start to be used with real energy costs that have to be met. The effect of these extra costs is to reduce profitability from its peak position. The point at which the free-issue resources are exhausted is the MSO point for the farm and this can be seen to coincide with the point of maximum profitability (measured as the 3rd contribution in Figure 8.4).

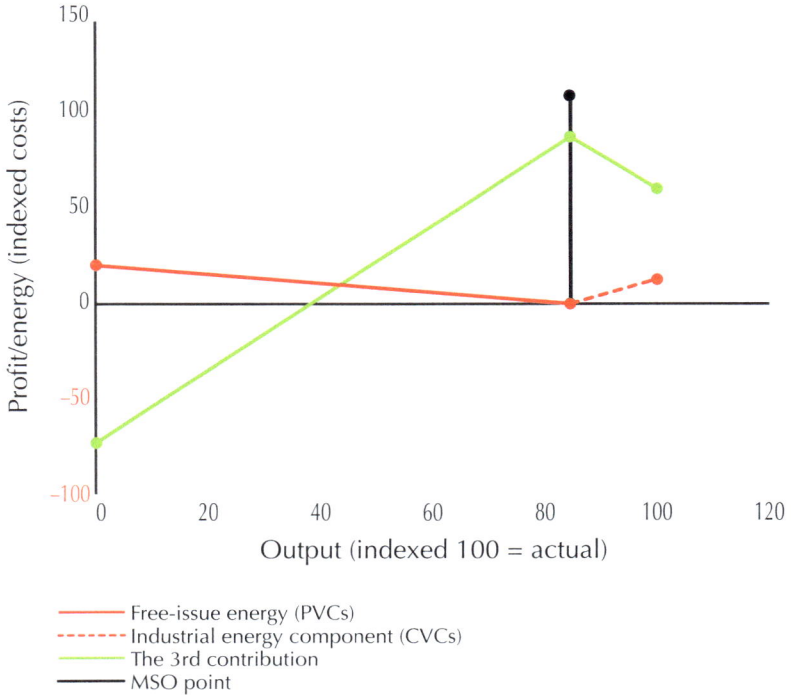

Figure 8.4 Energy balance diagram for a mixed livestock farm in the north of England.

The ultimate financial objective in any business is to produce a return on investment that beats deposit rates in a bank and rewards entrepreneurs for their enterprise and willingness to risk capital. Measuring this is encapsulated in a ROTA analysis (see Chapter 3). The ROTA chart for this farm business is shown in Figure 8.5.

The ROTA at the 2nd contribution level, on pre-support revenues is −8.21%. The addition of support payments is transformative. It improves the ROTA to 20.36%, which is well past the test rate of 15%. Moving to MSO will improve its ROTA further to 29.49%. A close inspection of the chart will show that the assets turn for the business is very low. Its best value is 0.62. Low asset turns are a common feature in farming but more could be done to win improvements on this count. The ROTA measure in the chart is for the business on a tenanted basis. Consequently, the value of the land is not in the calculations; this is the subject of the other (landowning) business where the imputed rent is the income that serves the business. The management accounts for the landowning business is shown in Figure 8.6.

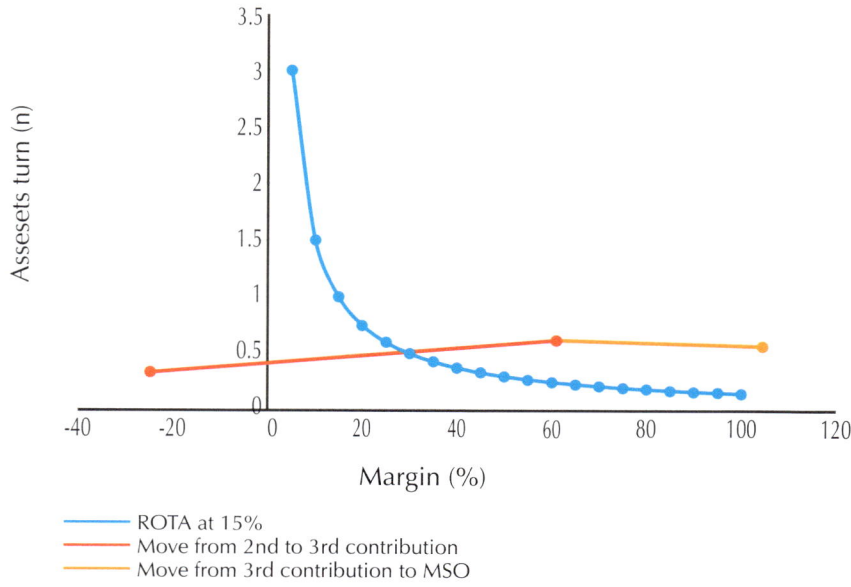

Figure 8.5 ROTA performance for a mixed livestock farm in the north of England.

Figure 8.6 Management accounts for landowning for a mixed livestock farm in the north of England.

Management accounts (as landowner)		£	%
Rental income (imputed)		25,000	
Land management costs		2,500	
Draw		0	
Net income		22,500	90.00
Land assets value		2,000,000	
Performance measures			
Margin	%		90.00
Assets turn	x	0.01	
ROTA	%		1.13

The ROTA performance of this part of the business is just 1.13% when the imputed rental is 1.25% of the land value.

However, livestock farming has to meet the burden of a heavy investment in livestock. In this case it is £120,000. The problem is not necessarily the size of the investment but its slow turn. The theoretical turn for stock that are taken from birth to full maturity is equivalent to 745 days-of-sale. This farm is working with an equivalent of 313 days-of-sale. For comparison, industry will target 45 days-of-sale, and live happily with 60. When these values reach 90 there is cause for concern. The impact of stock management realities and the costs of buildings and equipment all make for low asset turns. As indicated earlier, farms are not factories and will be unable to work their buildings, plant and machinery to the high degrees of utilisation common in industry. This all contributes to low asset turns.

The evolution and improvement of the unit costs of production in a business is a key indicator of its competitiveness. The calculation of unit costs is determined by the economic model being used. The usual practice is to treat all variable costs as a composite, as in the standard model of the firm. This produces a pattern of unit costs that continues to decrease with output but at an ever-decreasing rate. The MSO model produces quite a different profile, with a distinctive low point and ever increasing costs thereafter. The unit cost profiles for this farm are shown in Figure 8.7.

Figure 8.7 shows clearly that the MSO profile does not quite breach the competitive threshold on pre-support revenues. It also shows clearly the effect of pressing on beyond the MSO point; farmers following the standard model for future guidance will press on in the belief that all will come good at some point; farmers following the MSO model will be alerted to deteriorating prospects beyond the MSO point. These unit cost profiles are the only way to avoid some of the more subtle traps in marginal cost analyses.

The growth in the natural capital value of a farm business is an important measure of its success, commercially, environmentally and socially (see Chapter 6). On this farm, the PNC was being destroyed by the impact of CVCs and fixed costs and has ended up in a negative position. This farm is currently decapitalising its natural assets and this is shown in Figure 8.8.

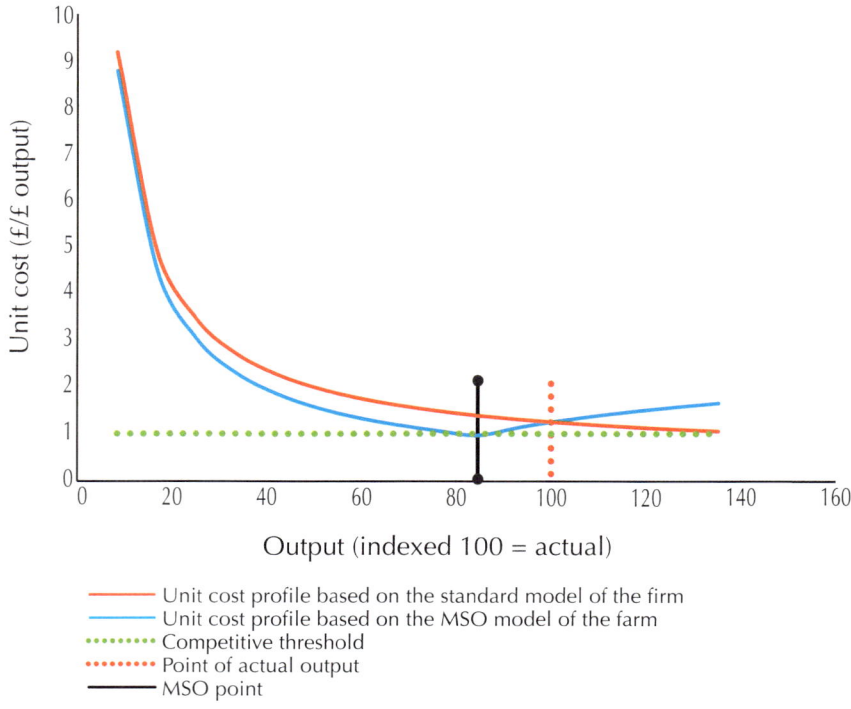

Figure 8.7 Unit cost profiles for a mixed livestock farm in the north of England.

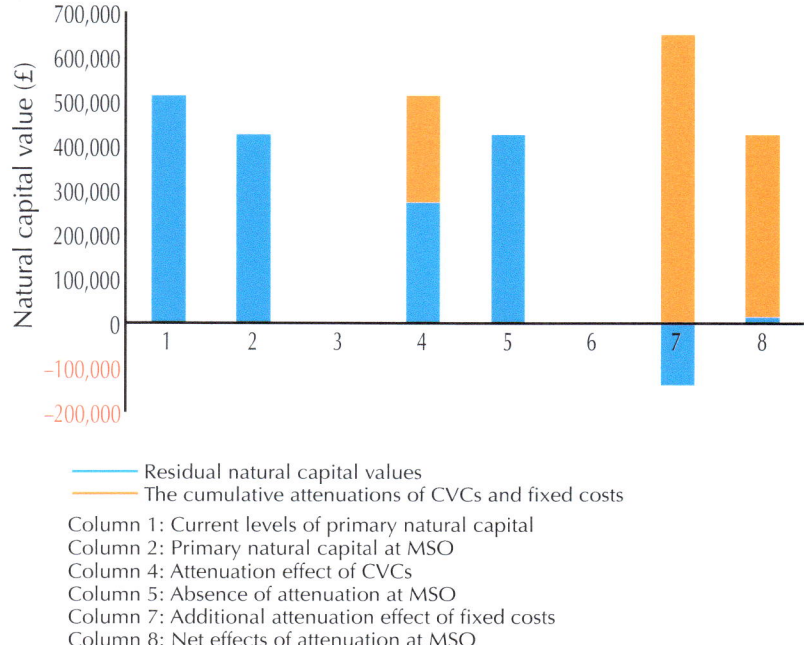

Residual natural capital values
The cumulative attenuations of CVCs and fixed costs
Column 1: Current levels of primary natural capital
Column 2: Primary natural capital at MSO
Column 4: Attenuation effect of CVCs
Column 5: Absence of attenuation at MSO
Column 7: Additional attenuation effect of fixed costs
Column 8: Net effects of attenuation at MSO

Figure 8.8 Progressive destruction of natural capital for a mixed livestock farm in the north of England.

The attenuation caused by CVCs will eventually overtake the income stream value attributable to Nature. When this happens, the business will have a sterile natural asset. The attenuation caused by the fixed costs is entirely commercial and when this is negative it becomes more difficult to deliver a real return on assets employed. On this farm, going to MSO would stop the drift to sterility and provide, just, an increase in the net natural capital value of the business.

8.3 CASE STUDY: SHEEP FARM, MID-WALES

Consider the performance of a farm of 265 ha in mid-Wales with 250 ewes. As the farmer owns the land, the business is split into two components: a landowning business, which charges an imputed rent to the farm operation and a farm business, which operates on a tenanted basis.

Its MSO ratio was established at 1.47 and its management accounts (on a tenanted basis), at actual and MSO, is set out in Figure 8.9. As the MSO ratio is greater than 1.00 the farm is operating below its MSO point. However, the presence of CVCs produce an inflexion point at 83% of its actual output level. This farm has, therefore a dual challenge; it should eliminate its CVCs and expand up to its MSO point to secure maximum profitability.

By expanding to MSO, together with the elimination of its CVCs, the 2nd contribution margin improves from −70.73% to +4.14%. Fixed costs are 134.06% of pre-support revenues largely the consequence of the small size of the farm and its low level of output. Even at MSO this only reduces to 91.06%. Support plays a massive part in this business providing 173.33% of pre-support revenues.

In this case, the farmer would like to expand (to MSO) and it would be advantageous to do so with an element of cattle. Many other farms in this position often find that expansion into cattle is prevented by grazing covenants. This is particularly true where grazing on the moor or commons in sensitive areas can be entirely prohibited.

In other cases where farms are operating below its MSO point farmers may be doing this consciously. When this is the case, the reason given is often that the farmer wants to 'restore the environment' or 'improve biodiversity'. As farm owners, it is legitimately their choice but it should be pointed out that not only is the nation being deprived of food output but that the path taken by Nature when taking-up the unused free-issue energy resources will be unpredictable and often disappointing and in none of these cases examined by the authors do the environmental payments received to promote biodiversity cover the lost profits of working at MSO.

Figure 8.9 Simplified management accounts for a sheep farm in mid-Wales.

Management accounts (as a tenant)	Actual	At MSO
SALES		
Lamb	30,000	
Beef	0	
Total sales	**30,000**	**44,167**
COSTS		
PVCs	2,000	2,000
Income stream from Nature	28,000	42,167
CVCs		
Concentrates	4,000	
Fertilisers	0	
Others	5,000	
Total variable costs	**11,000**	**2,000**
1st contribution	19,000	42,167
Fixed costs	40,219	40,219
2nd contribution	−21,219	1,948
Support payments	52,000	52,000
3rd contribution	30,781	53,948
Fixed assets	85,000	
Stock	26,000	
Assets employed	111,000	111,000
Performance measures		
Margin: 2nd contribution	−70.73%	4.41%
Assets turn: 2nd contribution	0.27	0.40
ROTA: 2nd contribution	−19.12%	1.75%
Margin: 3rd contribution	37.54%	56.10%
Assets turn: 3rd contribution	0.74	0.87
ROTA: 3rd contribution	27.73%	48.60%

The break-even chart for the business shows the size of the challenges for this farm. It is a very long way from its MSO point and its potential break-even point. This can be seen in Figure 8.10.

The underlying energy balance of this business is complex as a consequence of its working below its MSO point. The CVCs reduce profits beyond the inflexion point, but profits could grow in moving to the MSO point. This is illustrated in Figure 8.11.

A close inspection of the energy balance diagram will reveal that about half of the free-issue energy resources available on the farm are being unused. Nature, in appropriating this energy, had promoted considerable areas of bracken, as a dominant species in an unmanaged landscape, and this would have to be eradicated if the farm expanded its output.

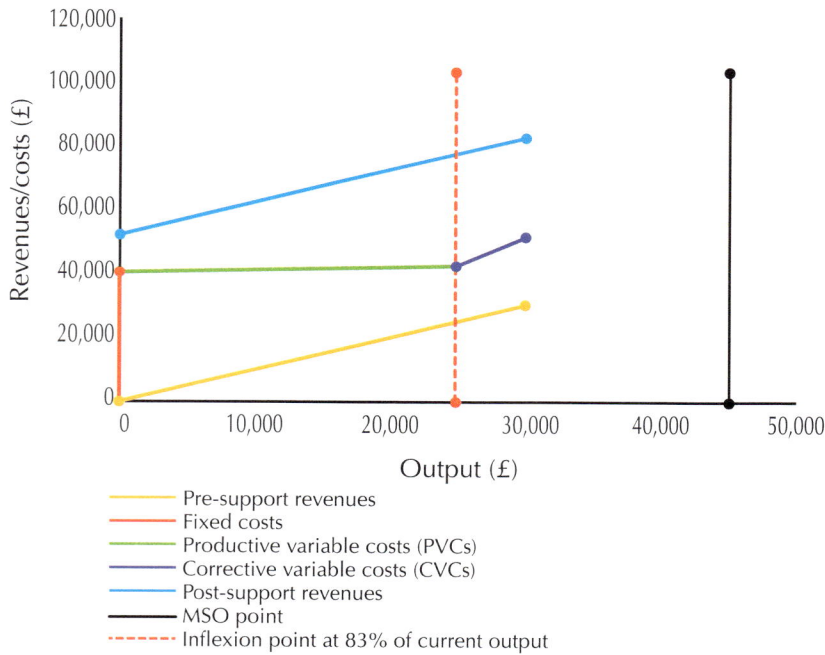

Pre-support revenues
Fixed costs
Productive variable costs (PVCs)
Corrective variable costs (CVCs)
Post-support revenues
MSO point
Inflexion point at 83% of current output

Figure 8.10 Break-even analysis for a sheep farm in mid-Wales.

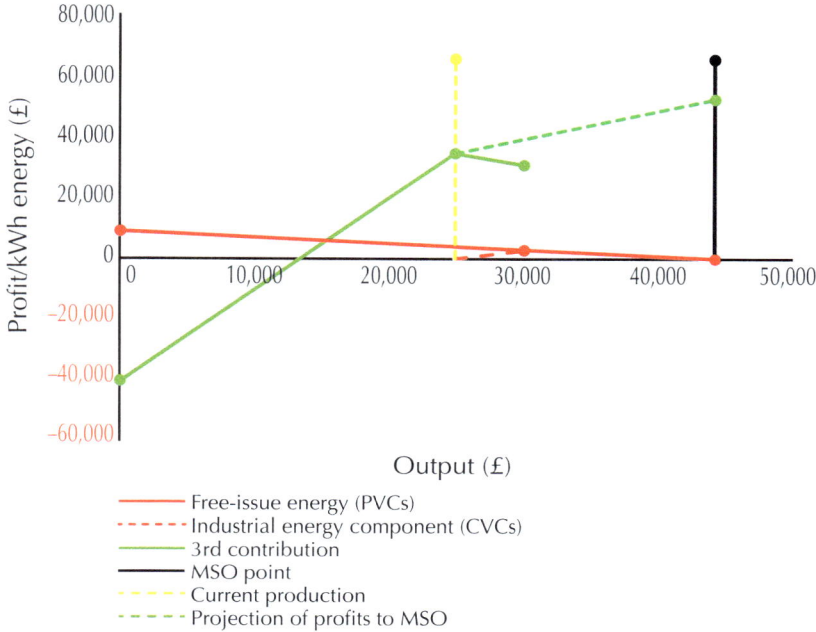

Figure 8.11 Energy balance diagram for a sheep farm in mid-Wales.

8.4 CASE STUDY: DAIRY FARM, WEST OF ENGLAND

Consider the performance of a dairy farm of 150 ha in the west of England with 250 milking cows. The dairy produced high quality milk at a premium price of £0.55/l to a milk company on contract. The product specifications in the contract had led the farmer to adopt a rearing regime with high levels of feed concentrates.

A dairy farm business will have a farm component, subject to the MSO model of the farm, and a factory component (the dairy parlour) subject to the standard model of the firm. The MSO ratio was established at 0.88 for the farm. Although the standard model does not exhibit an MSO effect per se, the farm MSO point will be reflected in the downstream output of the dairy and this was taken to be the MSO output of the dairy.

The two components of the business should be connected by a transfer price mechanism (see Section 3.7). This practice was not applied on the farm where the milking cows were presented on a free-issue basis to the dairy. The discipline of transfer pricing in situations like this will ensure that the farm can demonstrate its 'intrinsic' profitability. This always helps with better decision making. In carrying out an analysis of the business it was taken that the role of the farm was simply to supply cows for milking in the best possible condition. For this the farm was to be allowed to mark-up its costs to provide a suitable revenue line and a profitable margin for its role. The revenue line agreed for the farm, on this basis, was costs plus 25%. With a milk price of £0.55/l the effective transfer price from the farm turned out to be £0.25/l.

It can be argued that transfer prices are notional and when applied between two components of the same business why bother as it has no effect on the ultimate outcome. Apart from enabling better decision making for the farm, when fair and proper transfer prices are not used there is every likelihood that the downstream added-value part of the business will undercharge for its product. It is always easy to sell when produce is underpriced. Farms essentially produce commodity goods with little or no recognised differentiation and few opportunities to command a premium price. Added-value activities do have differentiation and the prospect of branding in many cases. A premium price at this point is much easier to win but the opportunities can be squandered when transfer price disciplines are not applied.

The management accounts for the business at actual and MSO are set out in Figure 8.12. The dairy farm was already very profitable. It had a very low level of fixed costs, at just under 20% of farm revenues and had a very low dependency on support payments. The farm component delivered a margin of 20.00% at the 2nd contribution level and the dairy component a corresponding 29.18%. Going to MSO would improve these to 72.12% and 64.94%, respectively. Beyond the MSO point on the farm the adverse CVCs leverage was £4.22 for every £1 of revenue earned. This can be seen clearly in the break-even analysis for the farm as shown in Figure 8.13.

The dairy component of the business, by contrast, follows the factory model of business (the standard model of the firm). Again, the activity was very profitable with an extremely low level of fixed costs, at just 6.5% of sales revenue, and a milk price of £0.55/l, that easily covered the transfer price of £0.25/l from the farm. At MSO levels on the farm, the transfer price for the

Figure 8.12 Simplified management accounts for a dairy farm in the west of England.

Management accounts	Farm		Factory (dairy)	
	Actual	**At MSO**	**Actual**	**At MSO**
SALES				
Farm income (from dairy)	621,094	546563		
Milk			1,077,314	948,036
Total sales	**621,094**	**546,563**	**1,077,314**	**948,036**
COSTS				
PVCs	32,000	32,000		
Income stream from Nature	589,094	514,563		
CVCs				
Concentrates	220,000			
Fertilisers	45,000			
Others	79,000			
Total variable costs	**376,000**	**32,000**	**692,000**	**261,375**
1st contribution	245,094	514,563	385,314	686,661
Fixed costs	120,000	120,000	71,000	71,000
2nd contribution	125,094	394,563	314,314	615,661
Support payments	25,000		25,000	
3rd contribution	150,094	394,563	339,314	615,661
Fixed assets	1,840,000	1,840,000	450,000	450,000
Stock	80,000	80,000		
Assets employed	1,920,000	1,920,000	450,000	450,000
Performance measures				
Margin: 2nd contribution	20.00	72.12	29.18	64.94
Assets turn: 2nd contribution	3.65	2.39	3.22	2.11
ROTA: 2nd contribution	73.07	69.85	231.88	136.81
Margin: 3rd contribution	23.10	29.18	73.34	64.94
Assets turn: 3rd contribution	3.80	2.39	3.36	2.11
ROTA: 3rd contribution	87.78	69.85	246.58	136.81

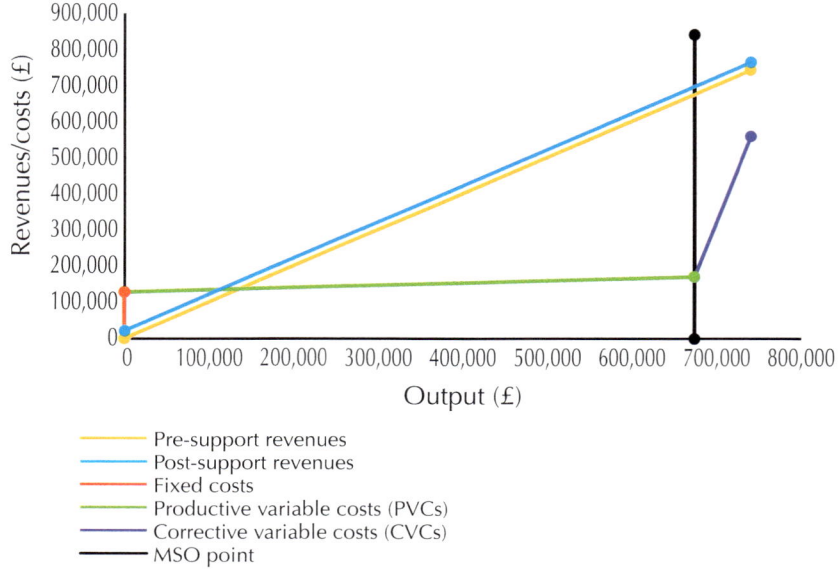

Figure 8.13 Break-even analysis for the farm component of a dairy farm in the west of England.

milk would be reduced to £0.11/l and the business profitability of the dairy component would be greatly improved as a result. This is shown in Figure 8.14.

The underlying energy balances on the farm are shown in Figure 8.15. The considerable use of concentrates and fertilisers meant that the incursion of these costs into profitability was particularly severe. Eliminating these implied a move towards an all-grass-fed herd and in this case, there was sufficient grass production capacity at the MSO point to allow this. The role of the feed concentrates in supporting the milk-count and the premium price presented a different challenge. The concentrates contained very high levels of phosphates and these were causing problems for the river authorities through the disposal of slurry on the farm. There was a prospect of sanctions being applied to manage this problem and the future of the dairy might be to produce milk at a lower price from an all-grass diet. A 10% price reduction to £0.50/l would reduce profits and the MSO 2nd contribution margins would decrease from a prospective 64.94% to 61.05%.

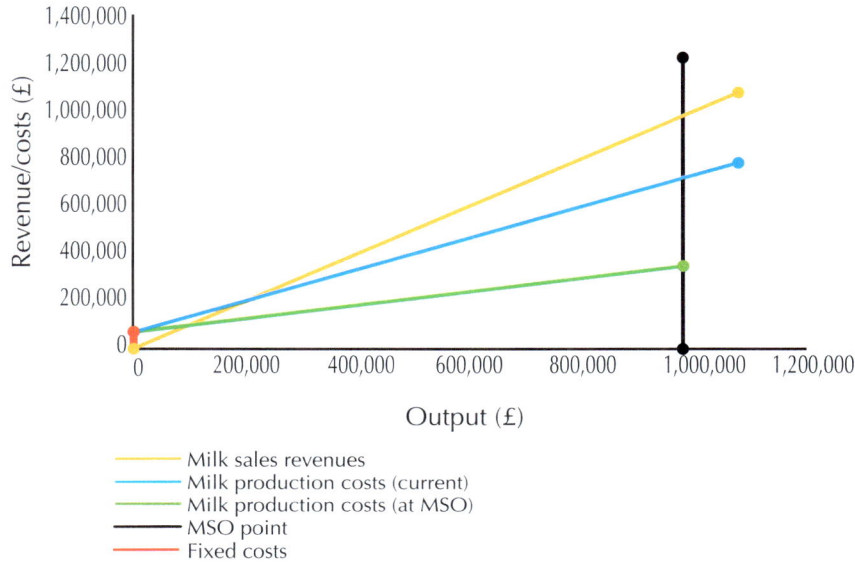

Figure 8.14 Break-even analysis for the factory component of a dairy farm in the west of England.

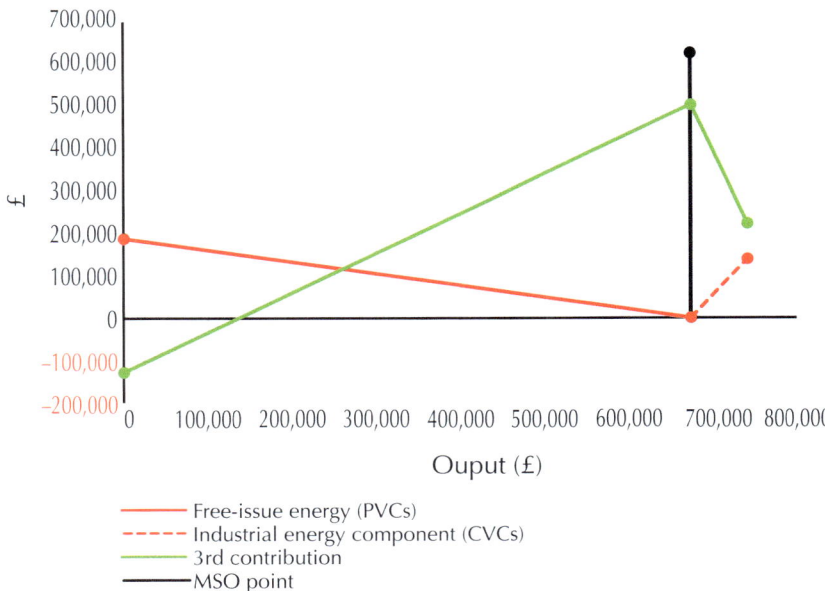

Figure 8.15 Energy balance diagram for a dairy farm in the west of England.

The two components of the business have very separate ROTA performances. The improvement in ROTA on the farm component in moving from its 2nd contribution to its 3rd contribution is small. This contrasts starkly with most non-dairy farms where this improvement is typically the greatest. This reflects the differences in the level of support. However, the transfer price discipline makes the farm very profitable in comparison to most others and this is the real value of added-value activities in the farming sector with its 'captured' market. Moving to MSO also added a further boost to performance.

The improvement to ROTA for the factory component of the business comes entirely from the improved productivity on the farm at MSO by way of a reduction in transfer prices. The effect on factory performance is considerable and parallels the farm improvement. This is shown in Fig 8.16.

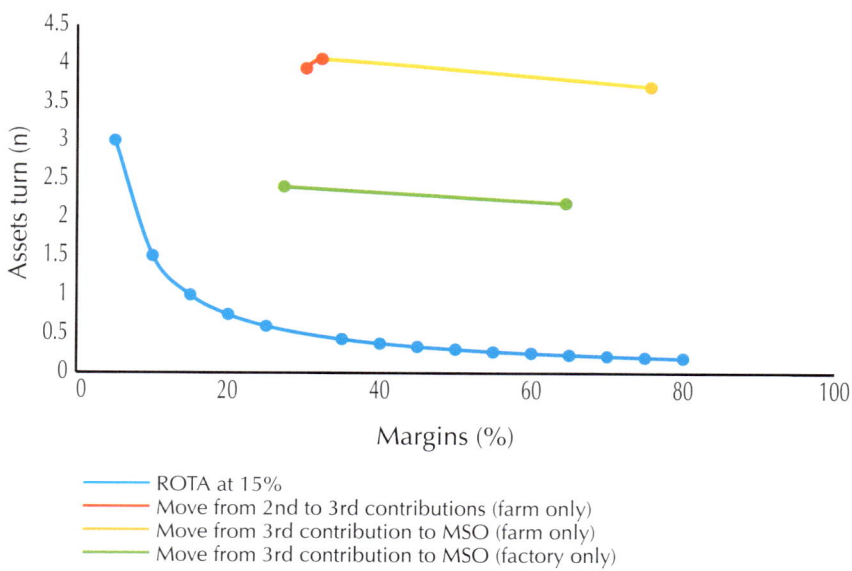

Figure 8.16 ROTA performance for a dairy farm in the west of England.

8.5 CASE STUDY: ARABLE FARM, EAST OF ENGLAND

Consider the performance of an arable farm of 1,400 ha growing cereals in the east of England. Its management accounts are shown in Figure 8.17

Figure 8.17 Simplified management accounts for an arable farm in the east of England.

Management accounts (as a tenant)	Actual	At MSO
SALES		
Cereals	1,600,000	1,326,295
Total sales	**1,600,000**	**1,326,295**
COSTS		
PVCs	147,690	147,690
Income stream from Nature	1,452,310	1,178,605
CVCs		
Fertilisers	350,000	
Others	376,690	
Total variable costs	**726,690**	**147,690**
1st contribution	873,310	1,178,605
Fixed costs	805,000	805,000
2nd contribution	68,310	373,605
Support payments	300,000	300,000
3rd contribution	368,310	673,605
Fixed assets	4,050,000	4,050,000
Stock	200,000	200,000
Assets employed	4,250,000	4,250,000
Performance measures		
Margin: 2nd contribution	5.53%	23.35%
Assets turn: 2nd contribution	0.94	0.78
ROTA: 2nd contribution	5.21%	18.22%
Margin: 3rd contribution	24.28%	50.79%
Assets turn: 3rd contribution	1.12	0.96
ROTA: 3rd contribution	27.14%	48.59%

The MSO ratio was established at 0.85 for the farm and in moving to its MSO point the 2nd contribution margin would increase from 5.53% to 23.35%. Fixed costs at 50.31% of pre-support revenues were on the high side for such a large business and support payments, although substantial, were just 18.75%.

The main component of its CVCs was fertiliser costs and these had been used heavily for many years. The farm had been seed-drilling for over 10 years. The adverse leverage of its CVCs was £2.04 per £1 revenue beyond the MSO point. This is illustrated in its break-even analysis shown in Figure 8.18.

Moving to its MSO point involved the complete cessation of fertiliser use, with a consequent reduction in yields. To replace some of the nitrogen con-

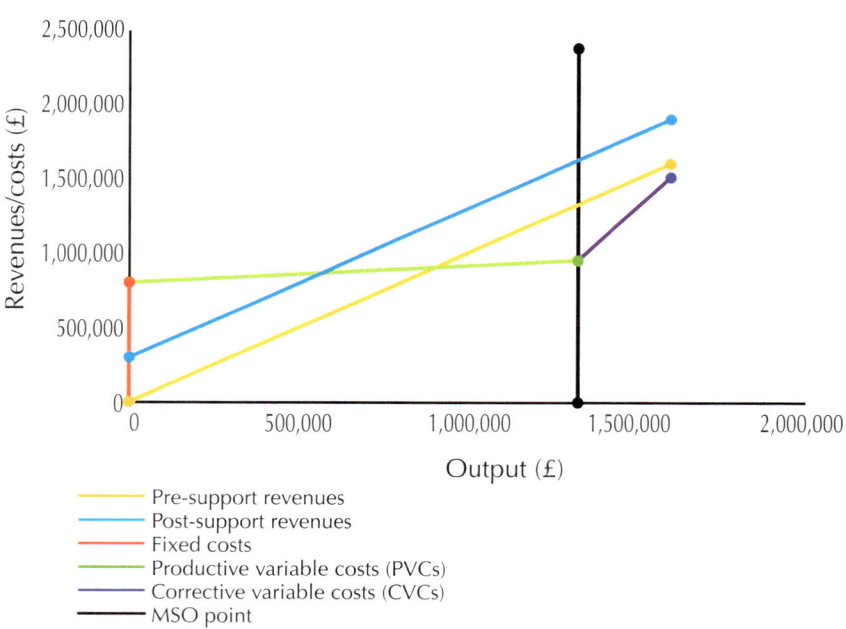

Figure 8.18 Break-even analysis for an arable farm the east of England.

tent lost in the absence of fertilisers the farm considered the re-introduction of some livestock. The farm comprised five different parcels and the re-introduction of livestock would force a consideration of some rotational policy. There had been no rotation of crops on the farm for some time. The biggest barrier to re-introducing livestock was the absence of suitable fencing. For a similar arable farm on the urban fringes the issue of re-introducing livestock, even with good fences, would have been impractical.

The energy balance on the farm clearly shows the high energy content of its CVCs and its impact on profitability. This is shown in Figure 8.19.

Despite the modest level of support, it was critical to the business in taking its ROTA performance beyond the 15% test line. Moving on to MSO would produce an even bigger improvement and this is shown in Figure 8.20.

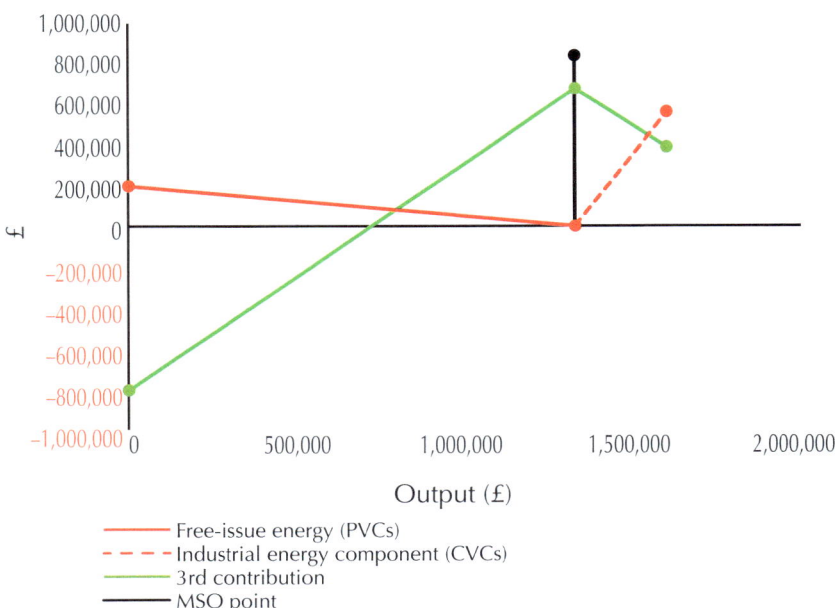

Figure 8.19 Energy balance diagram for an arable farm in the east of England.

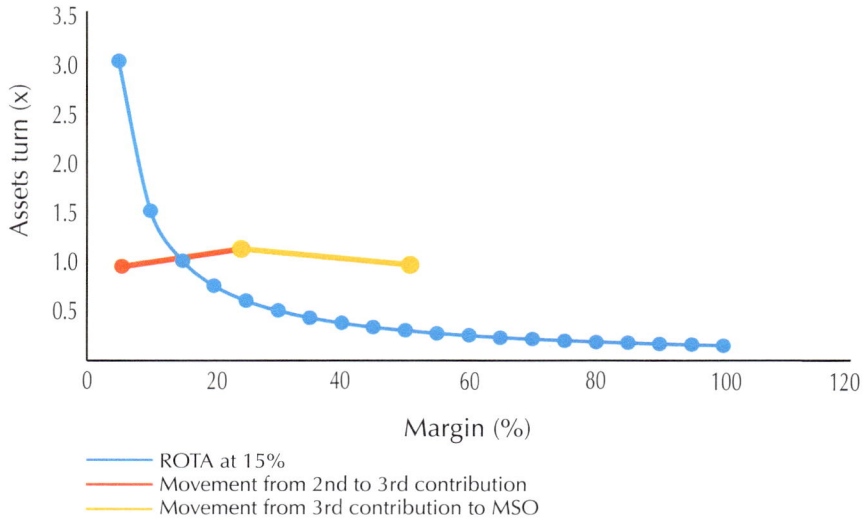

Figure 8.20 ROTA performance for an arable farm in the east of England.

8.6 ANCILLIARY CONSIDERATIONS

The disposal of slurry on a dairy farm presents challenging problems. It is still quite common to find slurry being spread onto the top of a field that runs down into a river course when there is little prospect of the main slurry content being absorbed in the pasture. Many farms now have invested in slurry storage, although few are covered facilities. The investments are substantial with a typical covered storage facility with a 9-month capacity costing £250,000. There is an interesting water management project in Northern Ireland where the slurry is run sequentially through three linked ponds before being discharged into a river course. The ponds are 200 m long by 3 m wide and cut into a hillside with a fall to the river course of 30 m. Its success is in its slow movement down each pond which optimises the deposition of undesirable elements in suspension. The water is clean on entry into the water course and from time to time the ponds are dredged for the silt which is then spread on the pasture. Slurry treatment is expensive, and the quality of its application is left very much with the individual sentiments of farmers.

Maize is being grown increasingly for animal food often on dairy farms. Currently, in 2024, about 180,000 ha in the UK are committed to this. While the maize crop can claim to be efficient in both water consumption and carbon sequestration it has a reputation for soil erosion and an appetite for herbicides. The more fundamental issue here, however, is the practice of growing crops for animal feed. It is profligate in energy conversion terms and it turns multi-gastric livestock into mono-gastric substitutes thereby by-passing the unique contribution these animals deliver.

Anaerobic digesters are a feature now on some farms. Some are run on maize and many are run on the waste from poultry farms. The poultry fed digesters can make commercial sense, but it must be reflected that the inputs may well come from very questionable poultry farm businesses. All crops grown for animal feed, including silage, and all digester outputs qualify as CVCs and as such will have an impact on the MSO point on a farm in relation to actual outputs.

8.7 DOWNSTREAM BUSINESSES

A downstream business is essentially a 'factory' at the farm gate that takes the output of the field and converts it into another, *added-value*, product. Dairy parlours are a special case; the animals, here, *participate* and return rather than being *committed* (that is, never to return).

A downstream business will follow the standard economic model of the firm; output must be driven hard, and equipment must run to high degrees of utilisation to be competitive. Where downstream businesses are attached, physically, to a farm property it is worth recognising that a successful downstream business may be much more profitable than its host farm and have an appetite that the farm cannot satisfy. Therefore, these factories should be sited on those parts of the farm property that can have independent road access. It is not uncommon to find some successful downstream businesses embedded in the farm building and inseparable from them. These businesses will be very difficult to sell-on when the circumstances may warrant.

Downstream businesses that are closely identified with a farm can fall into a trap by not being subjected to a transfer-price discipline. The principal objective of most downstream investments is often to exploit the intrinsic premium value in farm produce that is lost when it is sold at market prices on a commodity basis. If no transfer cost mechanism is applied, there is every prospect that the product will be underpriced in the marketplace. This of course makes it very easy to sell but it will be trading on the probable unprofitability of the farming element. In these situations, supplies should be transferred out of the farm and into the factory on a cost-plus basis. Even when the supplies priced this way are 'expensive' as long as the quality is there it will be much more possible to secure the required premium through the pricing of finished goods.

Meat products are a common downstream activity. Its essential prerequisite is to have animals slaughtered for a fee and returned to the farm for butchering. More will need to be done; stopping at this point puts a farm in competition with local butchers who will, typically, be more efficient, better-connected to consumers and have wider selections of product available. One route for farmers has been into ready meals by mail-order; others have entered the processed meat markets. Processed meats require very special skills to be differentiated but when these are acquired the prospects can be very rewarding.

Cheesemaking is an attractive option for some dairy businesses. As the quality and attributes of the product will be inseparable from the dairy herd, it is less problematic to have the factory element embedded in other farm buildings. Differentiation is the key, as with other farm produce.

Grain storage investments are quite different. The great advantage of storage, for up to 2 years of supply, is to take advantage of price movements. There is little added-value in these cases (and relatively little added-cost, too); it is a speculative game, which can be very rewarding and will require a trading mentality.

8.8 DIVERSIFIED BUSINESSES

A diversified business will not depend on any farm output; it will be driven more by the availability of farm properties for other uses. Many diversified businesses are extensions of some hobbies; equestrian ventures will often fall into this category.

Holiday accommodation is a growing area of interest, ranging from bed & breakfast to various forms of self-catering. These ventures are very separate from farming. A dilemma can arise when these diversified businesses, set up to enhance earning, outperform the farm business. Should the farm business then be retired in favour of the diversified business or not? In fact, there should be no real dilemma. No farm business should be retained simply because a diversified business can subsidise it. It should, and probably could, be made profitable in its own right.

8.9 STRATEGIC PERSPECTIVES

The construction of a business strategy has four fundamental steps.

The first step is product *differentiation*. A price premium will only be possible if a product is highly differentiated. In the absence of any differentiation the result is a commodity. Establishing a true differentiation is surprisingly difficult. Many product offerings will aspire to some differentiation but will fail in its delivery; other products will offer a differentiation that is more academic than real. What, for example, is so special about Welsh lamb? Or is it not so special at all? The old-wives-tale version was that it came ready flavoured from the rosemary covered fields of Radnorshire. Is this true? Either way, this is differentiation. In farming it could come from distinctive farm practices (provided these are real); the environment or the place; the breed; or a distinctive supply chain, etc. Downstream businesses are central to all this. Marketing professionals describe this step as 'selling the sizzle, not the sausage'.

The second step is to engineer *competitive advantage* into the business. Consumers expect both the best and the lowest-priced to come in the same package. Most competitive advantage comes from scale. However, as Nature is not scalable in a conventional way this will have a more limited impact in farming when compared to manufacturing industry. It is more likely in farming that true competitive advantage will come, in livestock businesses, from optimising the allocation of breeds to habitats. All-grass-fed livestock, capable of year-round living outdoors, on farm units above the minimum-economic threshold, working at MSO will take some beating. For cereal farming, the formula will comprise sunshine-hours, drainage and a self-sustaining nitrogen-cycle.

Competitive advantage in the 'factory' elements of the business will come from more traditional sources, such as scale and technical productivity.

The third step is to develop a stronger *grip* on the supply chain. In livestock farming this might involve avoiding sales to abattoirs, engaging in on-farm butchering, offering ready meals, etc. More grip will inevitably come from entering new downstream business opportunities. However, when grip is extended in this way, it must also be recognised that while the prospects for greater profitability might improve so the vulnerabilities associated with specialisations will increase business risks in general.

The fourth step is to extend *reach* into new markets. Lamb has been a popular product throughout the UK from time immemorial. With better international transport links and the recent emergence of some Islamic states as consumer economies, lamb supplies can reach out to these new opportunities. This must be done properly of course and, ideally, after building grip. Otherwise, such moves to extend reach would result in the long-distance shipment of live animals. This is not only detrimental to animal welfare but it would lose the commercial benefits of added-value.

All strategic business developments will inevitably require the re-structuring of organisations to cope with new roles and different scales. It is the failure to adapt organisations in the face of, otherwise, spectacular progress with differentiation, competitive advantage, grip, and reach that invariably cause the demise of many fast-growing hi-tech ventures.

CHAPTER 9
MSO FOR THE WIDER COMMUNITY

9.1 COMMUNITY CONTEXT

While the farm is an independent business it works within a landscape and is inextricably linked to other farms in its district. These farms operate as a local economic community. These communities have distinctive behavioural characteristics and an examination of economic performance from this perspective is important and has typically been neglected.

9.2 ECONOMIC CONTRIBUTION OF FARMS

The clusters of farms that comprise a local landscape will make a contribution to the local economy, of which, it forms part. This contribution comes in two parts.

- The net added-value from the cumulative value of the 2nd contributions delivered by each farm business.
- The net added-value from the cumulative value of the 3rd contribution delivered by each farm business.

The first of these reflects the commercial value of farm businesses at an operational level and the second the economic value of farm businesses after support payments and grants. The local economy will be driven, ultimately, by the second of these situations based on 3rd contributions.

However, when farm businesses are ranked, in order of their 2nd contribution, a distinctive pattern emerges when the cumulative effect is measured. This is illustrated in Figure 9.1 which examines the cumulative profitability of 17 farms in a local cluster when measured by 2nd contributions.

The first 8 farms in the cluster take the cumulative value to its maximum, as each makes a positive 2nd contribution; the next 9 farms take the cumulative value back to its minimum, as their negative 2nd contributions are added.

This phenomenon is known as the **hook-curve effect**. It exists for all communities and all economic sectors; only the scale changes.

Hook curves are the consequences of random behaviour. In this instance, the randomness reflects the impact of different farm sizes and differing levels of farmer competence. In a local community it may be assumed that climatic and geological differences will tend to be small. In an ideal community, the cumulative level of 2nd contributions would continue to rise, albeit at increasingly lower rates of improvement.

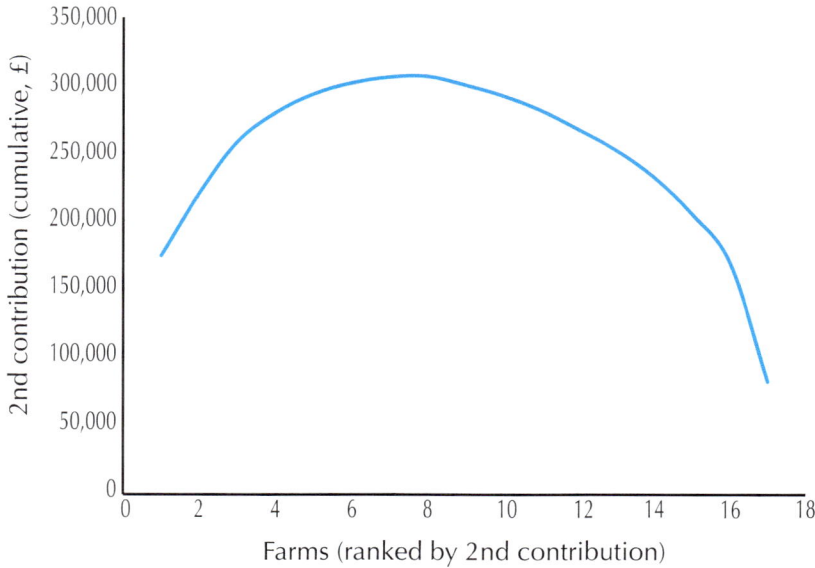

Figure 9.1 Community effect.

9.3 IMPORTANCE OF SUPPORT PAYMENTS

Support payments transform the net benefit of a farm cluster to its local economy. When the relevant 3rd contributions are plotted for the farms (ranked by their 2nd contributions) the hook-curve is improved considerably. In many cases, the impact of support payments will turn a negative cumulative 2nd contribution into positive 3rd contribution territory.

The impact of support payments on the same community of 17 farms can be seen in Figure 9.2. The cumulative profitability of the farms increases from less than £100,000 to over £600,000.

The blue line is the hook-curve representing the cumulative value of the 2nd contributions from the farms. It is the same as that in Figure 9.1-but the scale of the chart has flattened the curve. When support payments are added to the relevant 2nd contributions (representing the cumulative 3rd contributions), shown by the orange line, the contribution to the economy is transformed.

Support payments, which are often criticised for being spurious in their availability and application, do make crucial contributions toward a local economy.

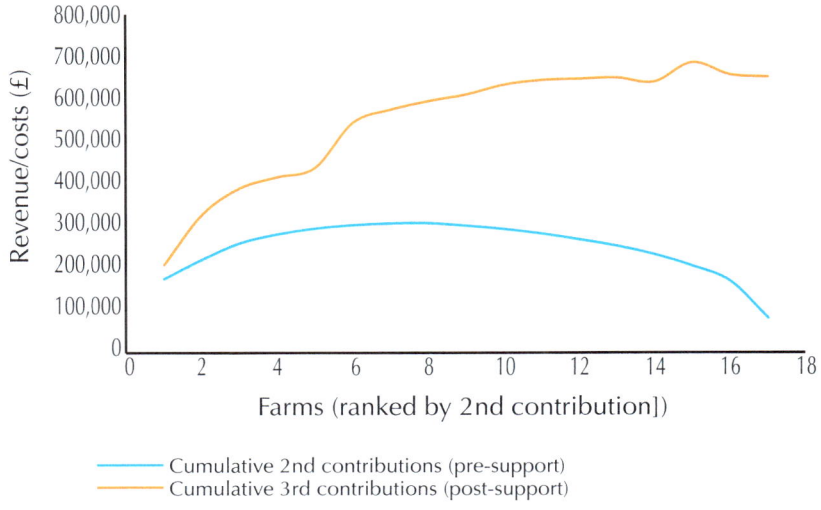

Cumulative 2nd contributions (pre-support)
Cumulative 3rd contributions (post-support)

Figure 9.2 Impact of support on farm communities.
Credit: Nethergill Associates Database.

9.4 COMMUNITY IMPACT OF WORKING AT MSO

Farms that move to MSO will enjoy an improved 2nd contribution. The impact of moving to MSO on a community, if all the farms were to do so, is illustrated in Figure 9.3.

The blue line is the hook-curve shown in Figure 9.1. Again, this is flattened by the scale of the chart. The impact of moving to MSO has been to turn the cumulative contribution to the local economy from just under £100,000 to just under £700,000. This is shown by the green line in Figure 9.3. For this cluster, moving to MSO would essentially replace the effect of support payments. As can be seen in Figure 9.4 where the impact of support payments (the orange line) can be compared with the impact of moving to MSO (the green line). Consequently, a move to MSO by all the farms in a local community will be a major factor in offsetting the removal of BPS payments, underscoring our earlier comments about the importance of achieving profitability on the 2nd contribution, which will improve farms' resilience in the face of changing/reducing support.

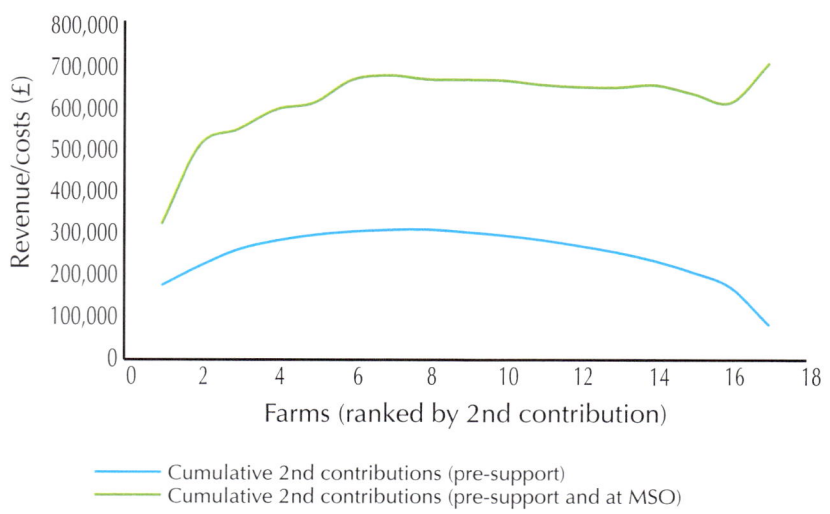

Figure 9.3 Impact of MSO on farm communities.

The full effect of moving to MSO and adding support is shown in Figure 9.5. It transforms the contribution to the local economy from just under £100,000 to just under £1.3 million.

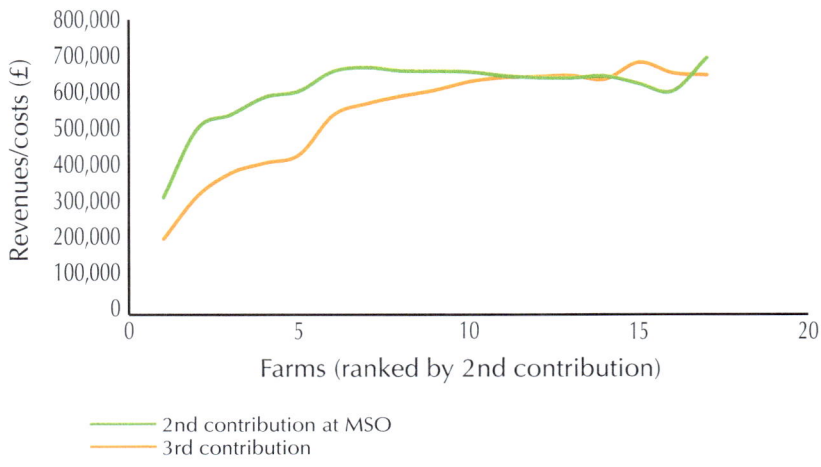

Figure 9.4 Relative impact of support payments and moving to MSO.

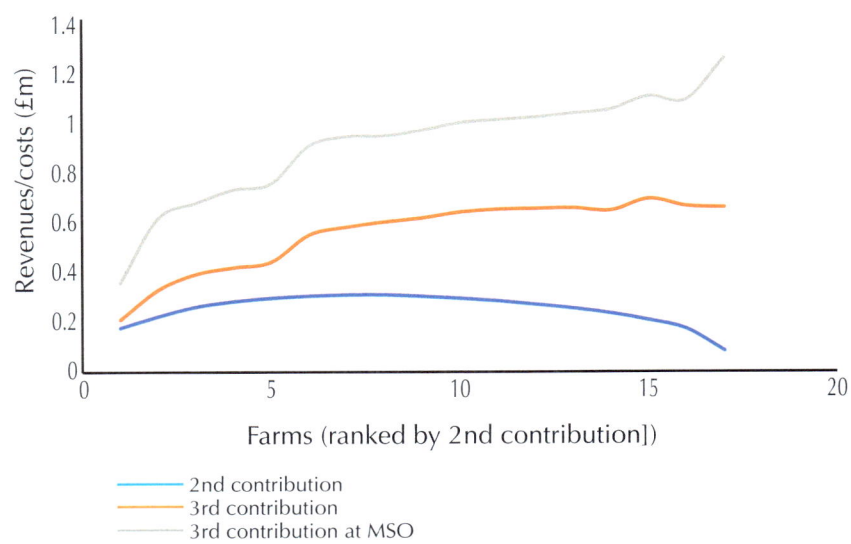

Figure 9.5 Full impact of support and MSO on a community.

9.5 NATURAL CAPITAL IN A FARM COMMUNITY

The natural capital of farms in a local economy can be treated in the same way and will produce another characteristic set of hook curves. This is illustrated in Figure 9.6 which relates to a community of 99 upland farms.

The blue line is the cumulative value the primary natural capital (PNC) of the group. This grows more slowly as smaller farms are added to the calculation. It will continue to grow as no farm should ever fail to cover its PVCs with its pre-support revenues.

When CVCs are taken into account there is an attenuation that reduces this to its secondary natural capital value. It can be seen that this reduces a PNC value of £76,000,000 to just £41,000,000. The CVCs in this case have destroyed 46% of the PNC value.

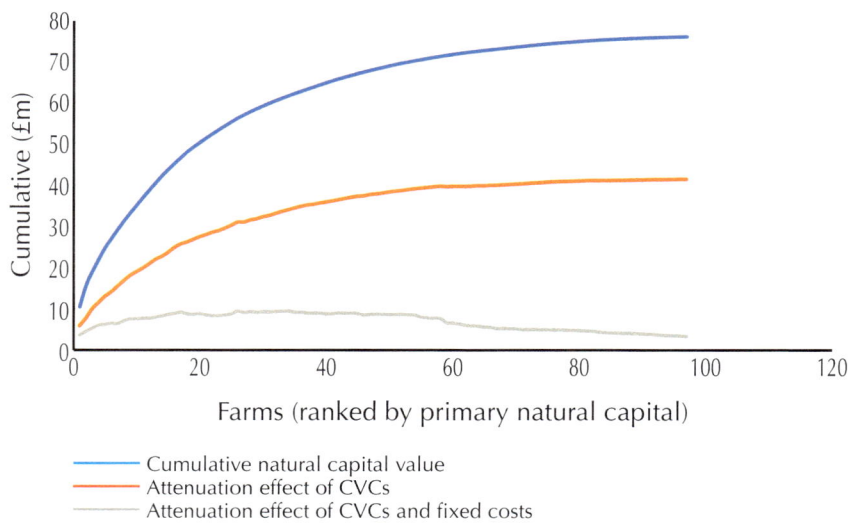

Figure 9.6 Natural capital in the community.
Credit: Nethergill Associates Database.

When fixed costs are taken into account further attenuation takes place and the tertiary natural capital value is reduced to just £3,000,000 representing a staggering loss of 96% of the PNC value.

This emphasises the fragile nature of natural capital and how easily it is destroyed by poor business performances. If farms were to move to MSO, there would be no attenuation between the primary and secondary evaluations.

Fixed costs are a major issue, especially in communities of small farms. This pattern of an underlying decapitalisation of the community landscape is masked by the role played by support payments. In effect, when the ultimate cumulative natural capital value for a community is negative the effect of support payments is to cover the cost of the destruction of natural capital at the courtesy of the taxpayer. This, alone, should be sufficient reason to review the need for and the role played by support payment schemes in the future.

9.6 ENERGY USE IN A FARM COMMUNITY

As set out in Chapter 5 farming can be examined through the lens of its being an energy producer and consumer. The consumption of energy on a farm comprises two, sequential, components:

- free-issue energy, such as sunlight, which drives the productive variable costs (PVCs)
- industrial energy, as embedded in chemicals and other industrial supplies, which drives the corrective variable costs (CVCs).

The cumulative values of the free-issue energy and costs of the industrial energy components can be established for a farm community, in a way that reflects the hook-curve analyses for contributions. The situation for a cluster of 17 farms (a different set from the cluster of 17 illustrated in Figures 9.1 and 9.3) is shown in Figure 9.7.

The cumulative 2nd contribution for the cluster is £90,000 (blue line). If all the farms in the cluster were to move to MSO this would improve to

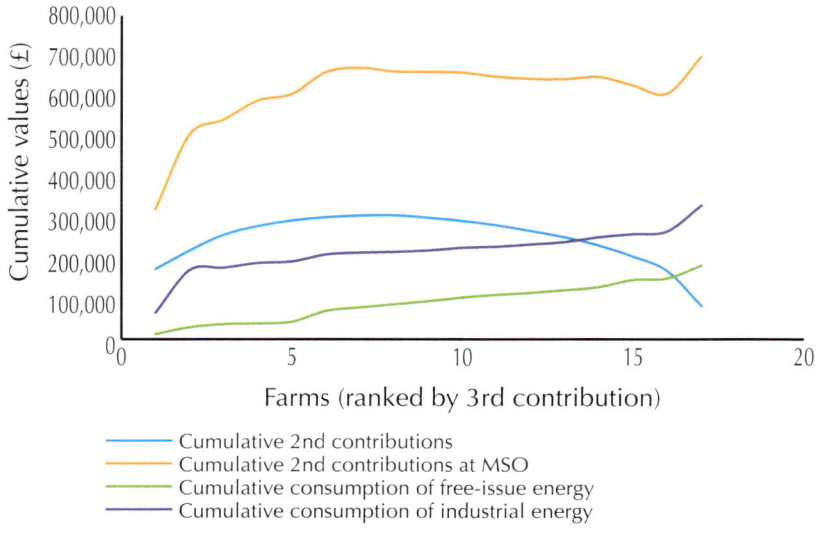

Figure 9.7 Accounting for effective energy consumption.

£700,000 (orange line). However, in delivering this economic contribution of £90,000 the cluster has used up free-issue energy (green line), with a notional value of £325,000, and industrial energy costing £180,000 (purple line). The conclusion is stark; the farm cluster has delivered less economic value to the economy than its consumption of industrial energy in this case, just 50% of the cost. This is another manifestation of Rule C (see Chapter 5).

Now consider the situation if all the farms worked at MSO. The contribution to the local economy would improve by 778% with much of this gain coming from the elimination of the industrial energy costs.

Additionally, if the cluster was to operate at MSO, and as this industrial energy content would be eliminated, the cluster could claim, justifiably, to be at, or as close as is possible to, net zero.

The case for this claim is simple. In our pre-industrial society, farming would have contributed to most of the CO_2 emissions of the nation. The growth in these emissions since then are the consequence of industrial output. If farming now eliminates all supplies with an industrial energy component from its business, it is essentially behaving as it did in a pre-industrial world. However, in practice, farming would not be in a position to eliminate the use of diesel fuel but as this is a vanishingly small component of emissions on a farm it can be neglected.

9.7 RESTRICTIVE COVENANTS AND ENVIRONMENTAL PRESCRIPTIONS

The use of common land for grazing has a long history and continues to play an important role today for many farms. Covenants, historically, were introduced to avoid over-grazing and to ensure a 'fair' distribution of benefits within a farming community. In recent times, new types of covenant have emerged, many of which impinge on farm profitability.

Covenants on moorland to maintain shooting estates will restrict the use of those moorlands for grazing; and covenants now being introduced by some commons committees have purely environmental or conservation objectives. Livestock, in many cases, are now regarded as a nuisance. It is worth observing, at all times, that the UK climate is superb for growing grass and any national agricultural strategy must recognise that its primary mission is to turn that grass into valuable food products. Not an acre should be wasted and not a kW of sunlight energy should be declined.

Consider a farm facing the negotiation of a new tenancy agreement with a major national institution. The institution had started to introduce new policies for its farm properties that reflected a significant change of direction in the pursuit of some new environmental objectives. In general, it wanted its tenant farmers to downsize in the cause of a better natural environment. When farms operate above their MSO point such a programme can be mutually beneficial.

In this case the farm was close to its MSO level of output but just under. Only about 7% of the farms in England and Wales operate below the MSO level. However, to be at MSO also requires the complete elimination of CVCs and this farm had significant CVC expenses. In ordinary circumstances it could become most profitable by eliminating its CVCs and expanding by 4%. As the farm was below MSO output levels a reduction in its concentrates and fertilisers would not have been accompanied by a loss of volume as it had the ability to produce all the grass it needed. Essentially, it had good prospects and an easier pathway available than most cases.

However, the tenancy renegotiations were turning on a renewal based on a significant reduction in stocking rates (entirely arbitrary) without any recognition by the institution of its impact on profitability. A concessionary rent was being offered to encourage the farmer to 'help the institution achieve its vision'.

In being pressed to reduce stocking rates to meet the new environmental objectives being adopted by the institution it would only seem fair to ask them to compensate for any reductions in profitability. This was established for them by making reference to the energy balance diagram for the business (Figure 9.8).

The current output index for the farm was 100. At this output the profitability of the farm was indexed at 45. If it was to move up to its MSO (output index 104) AND eliminate all its CVCs its profitability index would move to 58. If the farm was to be asked to reduce its stocking rates to 90% of current levels it would be operating on the dotted black line (output index 90) and its profitability would about 40. In other words, it would be sacrificing 18 on the profitability index (58 – 40).

If it was asked to reduce its stocking rates to 50% of current levels it would be operating on the purple line (output index 50) and its profitability index at this level is -2, representing a reduction of 60 from its MSO potential. If the institution was to compensate the farm for its

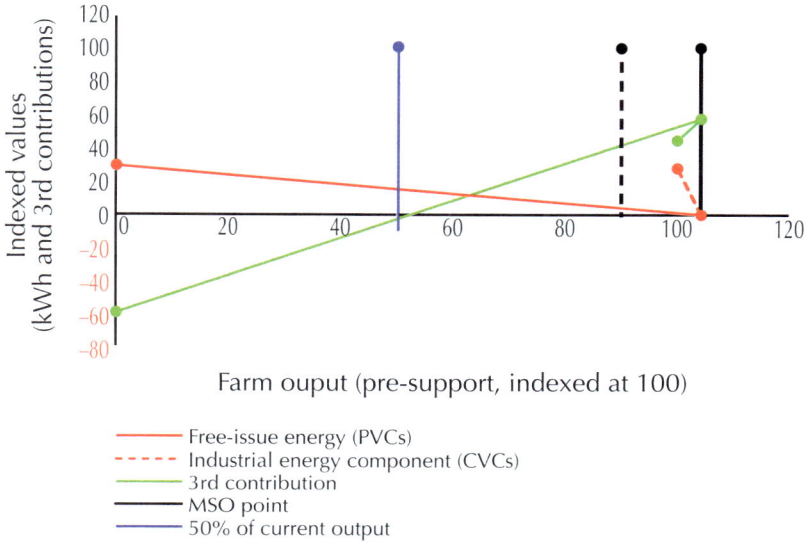

Figure 9.8 Energy balance diagram: case study farm.

loss of profitability in implementing their environmental schemes with these numbers, the farm would be simply neutral and no better off. The institution baulked at this point.

However, if the current tenancy came to an end and new tenants took over they, too, would have the same commercial realities to face (whether or not they might be aware of the numbers). The best solution all-round is to renew the current tenancy and let the farmer transition to MSO. For the institution it would be their least-cost option and at MSO the farm would exhibit an uncompromised pattern of biodiversity within its managed landscape. The pattern of biodiversity would not, strictly speaking, be maximised but it would be optimised for profitability. This is an entirely acceptable position and reflects the realities of sustainable farming. Also, at MSO, the farm business would have no industrial energy content imported in its inputs. The balance with natural energy sources will be a major contributor in the journey to a net zero position.

Furthermore, a profitable farm is a much better guarantee for the institution to charge a fair rent in the full expectation that it can be paid. As many institutions currently calculate their farm rents based on support payments, when the support payments change a new regime for rental calculations will have to be introduced.

9.8 WATER CAPTURE

Drinking-water resources are a key factor in an urbanised and industrialised economy. The greater part of these water supplies will have its origins on farmland. As part of its responsibility towards the managed landscape, farming has a role to play in providing clean water for our rivers and water courses. This is essentially a community issue for the water companies as it only takes one farm to compromise a whole water course with bad practices. At MSO a farm, in the absence of CVCs, will have a very much reduced impact, if any at all, on the quality of water courses.

The extensive use of agri-chemicals compromises water quality when application rates exceed take-up rates and surpluses run-off into water courses. The high protein content of some chemicals starves water courses of natural oxygen and other chemicals often introduce undesirable toxins, too. The dairy business is often dominated by the management challenges of slurry and despite significant investments in slurry storage too much is still dumped on fields only to leach away into water courses in wet weather. Many of the feed concentrates used in the dairy business contain high levels of phosphorus (P) and this is a particular problem in milk producing communities.

Hedges, which are often cited as major sources of biodiversity for better soil fertility, may have a more important role to play at the field-edges of water courses in filtering-out some of the chemicals that otherwise would run-off. As farmers are not, typically, compensated by water utilities for their water capture the investment in hedges to filter water run-off by the utilities may feature in the future.

CHAPTER 10
ADDRESSING THE FUTURE

10.1 THE MANAGED LANDSCAPE

The managed landscape, today, is not recognised universally as our de facto natural landscape. Yet, a return to the wilderness is just not possible, even in some limited form. There are simply no pathways back to some previous set of conditions. This is Rule D (see Chapter 5). Any attempts to visit radical change on the managed landscape will disturb the dynamic equilibrium to such an extent that it would prompt a plague of dominant species until a new point of natural balance became established. The energy to fuel these changes to the landscape could only come from those free-issue resources that are declined as farms withdraw from agricultural production. Any attempts to 'direct' Nature to a specific design through simple human endeavour will produce surprises rather than end in meeting its objectives. Some attractive results may well emerge, but these will likely be complete surprises.

The managed landscape in the future needs to be better recognised, more-nurtured and better-protected from adulteration through industrial chemicals. The landscape's role must be to produce food products or valuable fibres. Maintaining the managed landscape in its state of dynamic equilibrium with Nature, and 'nudging' it along a path of continuous improvement, will result in the least amount of effort and energy being expended. This attribute can only become more valuable in a world in which effective energy use is increasingly more critical.

Not all of this landscape can be made commercially productive. Elevation can be a big problem. Some of the landscape will also have an important non-commercial role, such as high blanket bogs in containing rainfall and steadily releasing the water. However, where the landscape has been modified successfully for food production over the centuries, such as the Fens, this should be retained. In these cases, it is more beneficial to society to retain the productive areas than to sacrifice them; an industrial society should be capable of designing solutions to emotive issues, such as destructive flooding.

10.2 FARMING

Farm businesses must return to a state of intrinsic profitability. Support payments should be phased out as expeditiously as possible and food prices at the farm gate will have to adjust. Political leadership is very sensitive to food prices, and its impact on voter sentiments, but the current patterns of support are probably more detrimental to national food security than the restoration of proper market prices as undersized farms and uncommercial farm practices are perpetuated.

Farm profitability could increase immediately as more farms move to MSO practices. No farm, with the exception of those undergoing a genuine transition, should be supported when it is unable to cover its variable costs and be profitable at the 2nd contribution. All farms should aspire to cover their variable and fixed costs or face the consequences of continuous decapitalisation. To mask these realities is to cheat society as cash drains away from the economy and business decapitalisations undermine a true market economy.

Farming, especially at a family-farm level, must become more business-oriented. When the sector has looked to industry for inspiration in the past it has focused more on industrialisation practices and the application of economies of scale to move to more intensive-working systems (which now look less and less appropriate); it has not examined the disciplines of more advanced forms of management control that come out of industrial practices in a comparable way.

On smaller farms, fixed costs are a particular problem. Often, the issue concerns plant and machinery. Farm equipment, apart from the tractor, will tend to have low levels of utilisation; the need for most specialist equipment will be confined to short-term situations and seasonal needs. When equipment is owned, it often deteriorates in storage for 9 months of the year. This creates a number of problems. It soon attracts heavy repair and maintenance expenses and it will be technically out-dated even after a few years and long before it ceases to function. Some farm clusters are adopting shared-ownership schemes for key equipment and, when these can be made to work (by avoiding the situation whereby all the participants need the equipment at the same time), the evidence points to a considerable improvement in profitability.

10.3 CASE STUDY: SHARING EQUIPMENT

In an upland farming district, where livestock farms were incurring fixed costs equivalent to an average of 65% of pre-support income, a group of four farms cooperated on equipment purchases. None of the farms in the district made a profit at the 2nd contribution when their fixed costs exceeded 40% of their pre-support income.

The situation for the four farms is summarised in Figure 10.1 The benefits of sharing reduced their fixed costs by between 68.93% and 17.50%: the smaller farms typically enjoyed the greater benefits and this can be seen in Figure 10.2.

Figure 10.1 Impact of equipment sharing on fixed costs.

Farm	Equipment sharing fixed cost (% pre-support sales)	No-sharing fixed cost (% pre-support sales)	Equipment sharing savings (% fixed costs)	Index of outputs
A	15.70	19.78	20.63	687.95
B	43.43	68.39	36.49	149.73
C	41.00	49.70	17.50	268.95
D	38.87	125.11	68.93	100.00

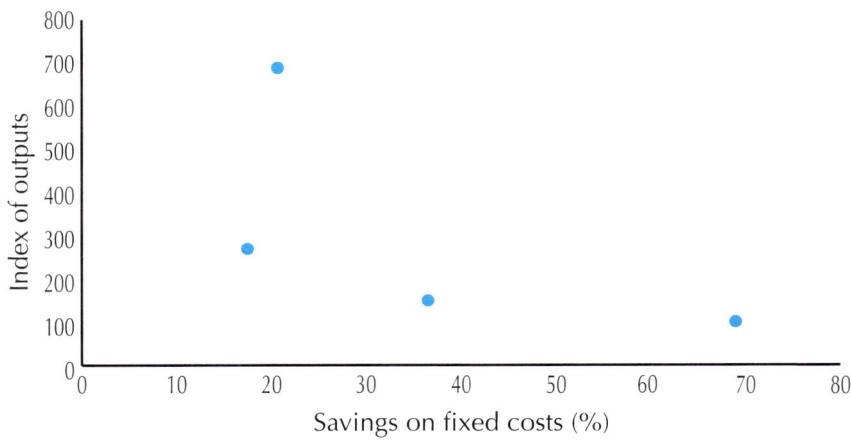

Figure 10.2 Impact of farm size on equipment sharing benefits.

10.4 AGGREGATION

As support payments are progressively withdrawn, and this is a most likely projection, many small farms will become a casualty. This will release significant areas of farmland for acquisition. It will provide an opportunity for many farms to expand by acquiring contiguous properties. A rental or leasing arrangement would put less stress on the new owner and, as the farmhouse asset is not a priority for an expanding business, it would allow the former farmers to maintain something of their former lifestyle.

Aggregation, in this fashion would expand the unit size of farm holdings, in general, and allow for the fixed costs in a farm business to be recovered more easily and competitively.

10.5 THE FOOD SUPPLY CHAIN

Grocery sales in the UK for 2023 were £171 billion. In contrast, the output from the agricultural sector of the UK in the same period was £14 billion (0.6% GDP). Of the grocery sales, over 80% was attributable to fewer than 20 supermarket chains.[9] An average supermarket chain, therefore, has a purchasing power equivalent to half the output value of the entire UK agricultural sector, which will comprise over 80,000 business units. This degree of mismatching ensures that farming, despite the vital role it plays in feeding the public, has little or no commercial leverage in the food supply chain. However, some new supply-chain models are beginning to emerge.

One of the more longstanding models is the 'flying herd', whereby a farm business avoids completely all the risks, and the lead-times, associated with rearing and buys in livestock for fattening or milking. These business enterprises can be very profitable for their owners. However, the real risk burdens are being taken by other farms closer to the origins of the supply chain. While the flying herd businesses can present themselves as organic or working at MSO this can mask the very different practices further up the supply chains.

Another is a more sophisticated version of the concept whereby investors have created a well-funded trading company to manage the early stock rearing months and then put out the livestock on grazing contracts to farmers with a surplus grazing capacity. These farmers buy in the livestock at an agreed price, set in £/kg, and then sell back at an agreed price, also in £/kg. The grazing farmers then have to worry only about costs. The trading company will typically settle sales invoices within a month and the farmers enjoy a much better cash-flow situation. The trading company, which may have an interest in special breeds, then pays for slaughter and markets a branded goods offering online. It has much merit, although, there are cases where the trading company has gone into liquidation and farmers have been left with animals to finish-off and sell at distress prices.

10.6 FOOD IMPORTS

It is generally believed that a heavily industrialised economy, such as the UK, will be unable to feed itself. This may never have been true, in general. Of course, some food products are simply un-obtainable from domestic sources, particularly exotic fruits and vegetables. Domestic supplies are often seasonal, and this has been an excuse fielded to justify imports in satisfying a year-round demand. Yet, little of the food available in stores is truly fresh; much of it will have passed through a refrigerated store. The critical issue in the food supply chain is simply price and the biggest determinant of price is often the cost of transport. In retrospect, this was the underlying factor in the years following the Napoleonic wars when it became possible to move grain from the North American prairies more cost effectively to our industrial centres than it was possible domestically, in the pre-railway era. With the use of refrigerated storage optimised for domestic output the UK may be far closer to feeding-itself than is generally perceived.

The UK has always espoused a preference for free trade internationally and a trade in food products would certainly help some of the less-industrialised nations of the world. However, food products are typically high-bulk and low-value, so it is somewhat bizarre to see vegetables air transported from Africa and South America to fill supermarket shelves. Air transport is only truly viable for high-value, low-bulk goods (the very opposite of food produce) and this practice is likely to attract more scrutiny in the future when sensitivities to air transportation become more acute.

Additionally, as some of the warmer and drier nations gear-up to produce fruit and vegetables for the industrial economies of the world they run the risk of depleting their own water table, as is now evident in East Africa and the Indian sub-continent. When the natural aquifers are fully depleted the whole nature of the landscape will change. This was the experience of farmers in North America when massive irrigation schemes were introduced into Kansas and when the cotton and oil industries of Texas depleted the Ogallala aquifer in the interwar period.

Some imported food products from advanced industrial economies come with very dubious credentials. American meat products, pumped-up with growth hormones, are ticking time bombs where the health of the nation is concerned; highly irradiated Dutch produce may have long shelf-lives to attract supermarkets but it will come at the expense of highly reduced nutritional values and possibly other, as yet un-recognised, dangers. In these cases, there is a possible role for non-tariff barriers to be invoked: food quality standards, foreign labour practices that are deemed to be exploitative and the true cost accounting of the imported carbon content could drive such initiatives.

10.7 ABATTOIRS

The time is long overdue to review the role and operation of abattoirs. Currently, abattoirs are the starting point in an industrialised supply chain for most meat products. Livestock are purchased almost entirely on prices set by simple demand and supply mechanisms. This is perfectly justifiable in a free market and it is, at least, a way of putting cash back into a farm business at an early stage in the supply process. However, two problems remain unresolved as a consequence.

Abattoirs are factories in every sense; they concentrate productive resources and they are driven by the need to run at high degrees of utilisation. This is not a good prescription for animal welfare when transportation is so stressful for animals and when the abattoir lairage often fails to insulate animals from a perception of doom.

Abattoirs also re-enforce the commodity status of livestock products through the price mechanism. While this prevails, there will be almost no opportunities for farmers to brand their goods in an attempt to achieve product differentiation (such as breed, heritage and countryside) in the pursuit of premium prices.

Mobile abattoir systems are now more credible, technically. Their adoption would allow farmers to pay a slaughtering fee and market their own differentiated meat products. Their economic viability will probably be driven by groups of farmers sharing the facilities with neighbouring farms.

By paying a fee for slaughter and working much more closely with butchering resources, farmers could control, in a much more effective way, one of the key factors in branding – provenance.

10.8 A ROLE FOR GOVERNMENT AND SUPPORT PAYMENTS

Government has been entangled in agriculture ever since the era of the Corn Laws. The repeal of the Corn Laws heralded a period of imperial free trade in agricultural goods with its attendant global supply chain in return for the export of manufactured goods. It made economic sense in a laissez-faire world on the grounds of comparative-advantage theory. It was more profitable to urbanise for manufacturing than retain resources in agriculture when imperial resources could be scaled up to meet most food requirements cost effectively. With control of the seas, food security was not an issue. This arrangement became vulnerable in the First World War when a convoy system had to be introduced to protect food supplies against enemy interception. Food security of supply has been vulnerable since then without being truly resolved. The thought, in the years after the Second World War, that this security might come from our integration into the European Community/European Union has not proved to be justified.

Redirecting domestic agriculture, following the First World War, to provide more was not a simple task. There was no equivalent of the great commodity trading-houses, that so characterised the imperial era, present on the domestic scene. To rectify this, state-controlled marketing boards were set up in the interwar years and, the last of them, the Milk Marketing Board was to survive over 50 years.

The Second World War repeated the exposures of the First World War and farm support payments were established to maximise domestic food production. It was a programme of production at all costs and it focused scientific attention on improving agricultural yields. The progress made in improving yields have been, technically, very impressive; however, it is more and more apparent that the technology has not been translated into better profitability.

Entry into the European Community took farming into its CAP and a very different world of state involvement. The CAP was little more, initially, than a simple device to ensure that German industry paid for the perpetuation of small French subsistence farmers, but, as with all attempts to manipulate markets, more and more of the agricultural sector had to be brought under control in order to maintain the system and cater for the 'specific needs' of others, such as Italian olive producers.

European practices of subsidy and state intervention became embedded with the CAP and remain today largely intact, conceptually, despite the withdrawal of the UK from the European Union. The CAP regime of BPS payments is being phased out and replaced, progressively, by ELMS. Essentially, area-based payments are being replaced by payments for public goods in the form of environmental benefits.

What are these goods? And what should be the role of support payments in general?

It is generally implied that support payments underpin security of supplies, but this causes two problems. Firstly, it tends to promote the maximisation of outputs. While this might seem to be perfectly logical, it has produced a focus on improvements to yields without considering the commercial implications. Yields can improve, almost indefinitely, if enough cost is thrown into the process. However, added-cost rarely translates into added-value in these circumstances. Secondly, support payments are not seen as funds to improve or develop farming practices; the monies are variously accepted as salary support, or price support or to cover losses. Apart from supporting different local economies by turning farm districts from being a net drain to being a net contributor, the degree of change that support payments can point to is negligible.

The concept of ELMS has all the hallmarks of environmental subsidies by the backdoor. Upland farmers could become park-keepers and lowland farmers could become little more than gardeners. It would seem to be the antithesis of true food supply security which can only come from an intrinsically profitable farming sector.

Government could simply remove all subsidies, allow food prices to adjust, and keep out dangerous foods and all other food imports other than by sea (that is all air-shipments and all long-haul road transport of milk, fruit and vegetables).

10.9 A ROLE FOR THE NATIONAL FARMERS' UNION (NFU)

A conventional trade union would see its role as one of free-collective bargaining with its' membership's employers. However, the farmers who comprise the membership of the NFU are usually employers not employees, which invalidates any concept of negotiating with an employer group. The main focus of attention by the NFU seems to be on influencing government policy particularly on issues such as food security (to maximise farm sector outputs) and support payments (to maintain farm incomes). The government seems happy to accommodate this. However, while this might occupy each party very satisfyingly the more important issue of developing a more sympathetic supply chain is neglected.

Farming needs better representation within the purchasing departments of the major retail supermarket chains. As with many industrial organisations, containing the cost of purchased supplies is a critical activity and two issues tend to emerge as common features.

- Price is important because it has a high leverage on profits.
- Transaction costs are also significant and there will always be a preference to deal with fewer and larger suppliers.

Mismanagement of these issues by the supermarkets could make them vulnerable. Lack of support for UK farm produce could jeopardise the domestic supply chain (and just at the time of increasing instability in international affairs). Reliance on fewer larger suppliers is just as risky as such companies tend to lose control of quality beyond a certain size as the scope of business activities becomes less manageable.

10.10 ENVIRONMENTALISM

Proper respect for the natural environment is entirely consistent with profitable farming. No business endeavour working with natural resources ever made a better contribution to the economy as a whole by despoiling the environment. Mis-managing farm resources simply results ultimately in decapitalisation. Farmers should not need to be *incentivised* to respect the environment; it should be abundantly clear that disrespecting the environment is simply not a long-term proposition.

Some aspects of farming, however, do result in environmental damage particularly where water courses are concerned. Run-off from fertilisers, pesticides and slurry are particular problems for rivers authorities and water utilities. Farmers have learned to farm livestock and crops over time; more recently many have learned to farm subsidies (to dubious benefit); perhaps it is opportune to learn to farm water and be part of its commercial footprint.

Concerns over pollution, in all aspects of public life, are driving campaigns to have the **polluter pays principle** adopted formally in the economy. Farming would be an early target for its application.

At MSO a farm will be in dynamic equilibrium with its managed landscape and deliver:

- maximum profitability, at the 2nd contribution level
- minimum energy footprints
- optimised patterns of biodiversity, for the farm in its environment.

The biodiversity is optimised in the sense that, at MSO, there is nothing being done on the farm that should spoil the environment. This is sufficient in itself to protect and maintain the natural environment.

10.11 CONSERVATIONISM

Conservationism is essentially a modern phenomenon; it is not easy to define but it commands great support and is seen by most elite and middle-income societies as a 'good thing'. However, it often strays quickly into territory that impedes businesses and genuine progress. All change has an impact, but the needs of an industrialised economy cannot always be relegated to second place on marginal and often spurious arguments about the relative value of Nature and economic activity (which is work often dressed-up pejoratively as profit).

However, the general public often express a sense of ownership when it comes to the common landscape despite the land in question being the private property of others. In the late 1980s there was a project to build a new freight railway from the Midlands to the Channel Coast. When the company sought to put a line of navigation through open and often un-productive countryside it was always vehemently resisted, even though the landowners were usually very open to an arrangement. When the company sought to put a line of navigation through some dense suburban housing in South London the only objections came from those property owners who had not been approached to sell-up (being outside the lines). The countryside was clearly part of their 'estate' but a neighbour's house was none of their concern.

Having a romanticised picture of the landscape is most unhelpful for commercial farming. A pretty landscape is just that. A working landscape is a thing of much greater beauty and certainly has more utility.

The landscape should remain as a contract (of duty) between its owners and Nature.

10.12 CLIMATE CHANGE

Today, nothing would seem to threaten the managed landscape more than some of the concerns around climate change. Farming not only utilises free-issue resources such as sunlight, but its success also depends on the enlightened management of the nitrogen cycle, the carbon cycle and the water cycle. Grassland and woodland comprise the default natural vegetation in the UK. Our climate is particularly well-suited to growing grass and ruminants are very efficient in converting that grass into edible protein. Inevitably, these animals, and the people who tend them, will exhale CO_2, and expire NH_4.

In the British economy of 1750, farming would have been contributing almost 100% to the carbon emissions of the day without any cause for concern about global temperatures. While the output of the farming sector has grown in real terms since 1750 its current 40% contribution must still broadly underpin the 15°C baseline temperature. In other words, all the increase in global temperatures must be attributable entirely to industrial activity. In 1750 British carbon emissions were believed to be 9.3 million tonnes per annum. In 2022 this figure was 318.6 million tonnes for the UK.[10] It is generally regarded that 40% of this is attributable to agricultural activity and this would amount to 127.4 million tonnes. The difference between the 1750 value for farming and the 2024 value will be due to two factors: firstly, the increase in output as the British economy expanded and secondly, the industrial intensification of agriculture from the middle of the nineteenth century. This industrialisation is encapsulated in the use of CVCs. If farming was to eliminate its CVCs, and operate at its MSO point, it could claim to be in balance with natural energy sources (i.e. at net zero); its grazing livestock has no justifiable role to play in the control of emissions.

The CVCs in farming, which comprise largely the industrial products of the agri-chemicals sector of the economy, are the repositories of the principal industrial energy content in the sector. With their elimination, farming will be left with little more than the need to reduce its consumption of diesel fuel to be close to its pre-industrial contributions to carbon emissions.

Furthermore, while governments respond to the popular demand to the need to reduce emissions none abandons the desire for continued economic growth. These objectives are, in fact, mutually exclusive. The rapid economic growth experienced since 1750 has been exclusively powered by the energy resources from fossil fuels. These fuels have been powerful, relatively cheap and very convenient to use. Their removal from the economies of the world may bring future economic depression and severe reductions in standards of living worldwide unless societies develop new strategies to cope with this situation. The economies of the developed world are all predicated on growth and if this growth is to be maintained, the world will have to live with climate change and recognise the importance of fossil fuels; this is also likely to be the pragmatic outcome, anyway, as our collective ability to control the forces of Nature involved is simply inadequate and never can be anything but inadequate. If anything, the undeveloped world is even more committed to growth as a way forward. We should try to avoid a Canute syndrome taking hold in Western middle-class societies.

Undoubtedly, CO_2 levels are now a great problem. Concentrations are increasing at about 10.5 parts per million by volume (ppmv) per decade. These concentrations levels will rise to 392 ppmv by 2030, and 403 ppmv by 2040.[11] This would make CO_2 only 0.04% of the atmosphere but its impact is very much larger. At this rate of increased concentration, the ambient global temperature will rise from 15°C to 17.5°C and sea levels will rise by 800 mm per century as glacier water is melted.

Human intervention to address this in the form of statutory de-carbonisation programmes come in two guises. Firstly, by reducing the consumption of fossil fuels, which will ultimately be manifested in population reductions as the world's economies go into reverse in the absence of growth. Secondly, by pursuing programmes to create new carbon sinks, such as new forests.

Nature will also continue to intervene. Excess carbon dioxide is a fuel for plant life. Unfortunately, the growth in plant life is outpaced by the combustion of fossil fuels and the opposing forces work on very different timescales. As global temperatures rise so will the evaporation rates of water from the oceans. This will result in a growing level of cloud cover and more rainfall. The extra rain will wash some of the surplus carbon dioxide into the oceans to be sequestrated there; more encouragingly, the extra cloud cover could change the albedo ratio. The current albedo ratio is taken to be 0.30 and comprises a mix of different components. For example, asphalt is 0.04; grass is 0.25; fresh snow is 0.85; and cloud is 0.65. At an albedo ratio of 0.30 the ambient temperature of Earth, before greenhouse gas effects, is −20°C with greenhouse gases responsible for adding 35°C. If the planet was entirely covered with cloud, with an albedo ratio of 0.65, the ambient temperature of Earth would be −59°C, a reduction of 39°C.[12] This will not happen, of course, but a small change in global cloud cover could reverse the global warming effect of carbon dioxide. The issue is very finely balanced.

Pragmatically, farming in the UK should position itself to exploit the new benefits that could come from a warmer, wetter, but still temperate climate.

10.13 POLITICAL ISSUES

De-carbonisation in the economy will have to be institutionalised if it is to work. This has constitutional implications.

All capital assets have to be maintained. This involves repair work, to maintain full working-order; remedial work, to correct for neglect; and enhancement, to adapt for changing circumstances. If the costs of maintenance are not met out of current revenues the result will be either decapitalisation, if there is total neglect, or passing the burden onto future generations, if maintenance is funded through debt. Meeting the cost out of current revenues is consistent with the **intergenerational obligation**, which seeks to ensure that assets are passed on from one generation to the next in an ever increasing better-order.

The intergenerational obligation gives rise to the **precautionary principle**, which seeks to ensure that the capital assets base is not being continuously destroyed through neglect.

To ensure that costs are properly assigned within the economy a further discipline needs to be introduced. Bad practices in one part of the economy will result in others having to meet the burden of corrective action. For example, farming has a reputation for failing to control run-off into water courses, which adds to the cost of providing clean water and drinking water for the community. This sort of behaviour has inspired the **polluter pays principle** as a way of containing and reducing such problems with benefits all-round.

Carbon taxes, which are widely touted as inevitable in the future, would be an example of the polluter pays principle.

None of this new behaviour will emerge from the classical laissez-faire/invisible hand mechanisms of the free market. If it is to happen, there will have to be not just legislative initiatives but fundamental constitutional changes.

10.14 CARBON TRADING

As the BPS programme is being run down, particularly in England, some small farms are looking for salvation in selling carbon offsets. Great caution is needed here.

Carbon trading does nothing to reduce carbon emissions; it merely 'green-washes' those business activities with high carbon emissions who wish to avoid a proper re-configuration. Some farms are being offered attractive one-off payments by large corporate enterprises in return for adopting more 'carbon-friendly' protocols on the farm. In the small print, the application of these protocols will require constant review, and often improvements, for over 100 years. From a fundamental business standpoint, it is quite untenable to book a single payment which carries with it the obligation to spend money continuously into the future. This was the feature that lay behind the collapse of the American energy company ENRON at the turn of the century where long-term supply contracts were booked as income in the year they were signed leaving producers with the costs of supply for the remaining term of the contracts.

Carbon reduction models are sometimes no less dangerous. Most dairy companies seem to have bespoke carbon reduction models. In return for complying with the objectives of these models, dairy farmers can attract a premium price. The principal objective of the dairy companies is usually to reduce the consumption of diesel fuel. This is quite simple to achieve if the number of collections can be reduced without sacrificing volumes. Dairy farmers are therefore encouraged to increase outputs of milk per cow. However, this is diametrically opposed to the concept of optimising the trade-off between milk outputs per cow and its number of lactation cycles. This optimisation favours lower-outputs, longer lactation cycles and all grass-fed dairy regimes.

10.15 THE NATURAL DIVISIONS OF THE UK

The UK can be divided into 32 natural regions, each of which will have typical and distinctive set of farming conditions. The full scheme can be found in the appendix section.

Standard farm types need to be defined for each of these divisions as a first step in building a national framework of standards for the industry. These standards should reflect the methodologies of industrial engineering practices as applied to industry. This approach would provide a set of synthetics from which any farm operation in any configuration can be replicated. This will not be either a simple process or be a quick one; it will be expensive and unless a professional company undertakes the work as a commercial project (which was the case for manufacturing industries in the United States) a sector-wide initiative for farming would need to be funded.

Until this happens, the best that can be achieved is to use some typical farm types for each of the 32 natural divisions and establish some *best practice models*.

10.16 RESILIENCE

Farming exists in a world of uncertainty. The weather is unpredictable and the sector is exposed to the shocks of changes in political sentiments as governments change and world events affect the supplies of goods and services.

Resilience is a critical attribute in mitigating uncertainties and shocks. In farming it is a function of two complementary parameters:

- the point at which break even on a farm is achieved by its pre-support revenues on its fixed costs and its PVCs. The earlier this occurs in volume terms the more resilient the business will be to changes in sales volume (for whatever reason).
- the potential for the farm to benefit from moving to its MSO point.

A resilience index (RI) can be defined as follows:

$$RI = (1-p)*100$$

where p is the ratio of break-even point to output.

In this formulation, the RI can never be more than 100 and the higher the RI the more resilient the business will be. When the potential to move to MSO is included, in a composite resilience index (CRI), the formulation will be modified to:

$$CRI = RI*q$$

where q is the ratio of MSO to actual output.

Therefore, if a target CRI of 30 was taken to be an objective, in a fashion similar to ROTA considerations, there would be an infinite number of combinations of RI and q to deliver this. This is demonstrated in Figure 10.3.

An example of this analysis as applied to a small community of seven farms is set out in Figure 10.4 One farm was working well below its MSO point ($q > 1$) and this farm had a negative resilience. The other farms in this community all beat the CRI = 30 threshold.

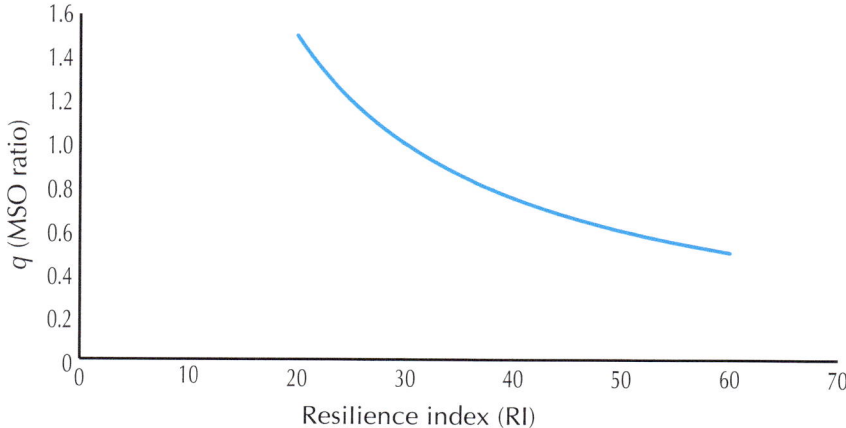

Figure 10.3 CRI = 30 diagram.

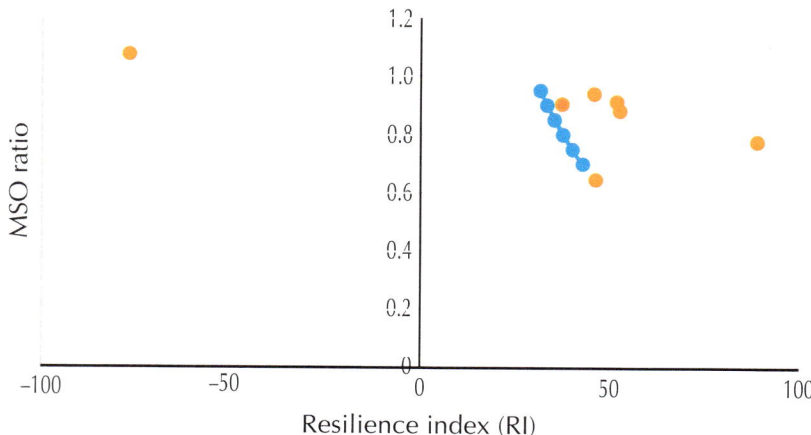

Figure 10.4 Farm resilience in the community.

Chapter 11
AN OUTLINE PROGRAMME FOR BUSINESS DEVELOPMENT

11.1 HIERARCHY OF PERFORMANCE TESTS

The **first objective** of a business is to ensure that it is not losing cash on every transaction. Such businesses are simply unviable. The simple test is:

Revenue from sales *less* variable costs must be positive.

Surprisingly, about 18% of all farm businesses in the UK will fail this test and still survive. When this happens, support payments and other grants and subsidies will more than offset the working deficits.

The **second objective** of a business is to ensure that it is not decapitalising. Such businesses will eventually become bankrupt when all its capital assets are exhausted. The simple test is:

Revenue from sales *less* variable and fixed costs must be positive.

Fewer than 20% of all farm businesses in the UK will, at the time of writing in 2024, pass this test. Again, support payments and other grants and subsidies come to the rescue. As most farm businesses are partnerships of some form the fixed costs incurred in this test will not include any drawings.

Increasingly, farm businesses are being invited to participate in landscape recovery schemes where payments are available from the state to implement desirable environmental protocols under the banner of **public payments for public goods**. It could be argued that participation in such schemes should qualify these payments to be considered as proper sales revenue. If this convention was adopted it would be less demanding to meet the first and second objectives above. However, it must be remembered that to do so would be an acknowledgement that the business would become a hybrid of farming and environment management. As all public payment schemes cannot be relied on to be either consistent or permanent this position is not favoured by the authors.

The **third objective** of a business is to allow its partners or owners to draw a fair compensation for their work. What constitutes a fair draw is subjective and a judgement call but a starting reference point might be:

50% above average earnings.

Farmers tend to work hard, put in long hours and be modest in their expectations. Even at 50% above average earnings the likely hourly-rate equivalent will not be too attractive to observers outside the farming sector.

The **fourth objective** of a business is to deliver a competitive return on the capital employed in the business (ROTA). Capital invested in a risk-free instrument, such as deposits in a building society, will attract, say, 5% pa compound interest. As businesses are not risk free it is only fair to expect a premium on this rate for the effort and enterprise involved. Again, a fair premium is a subjective issue but a starting point might be 10% resulting in:

15% ROTA

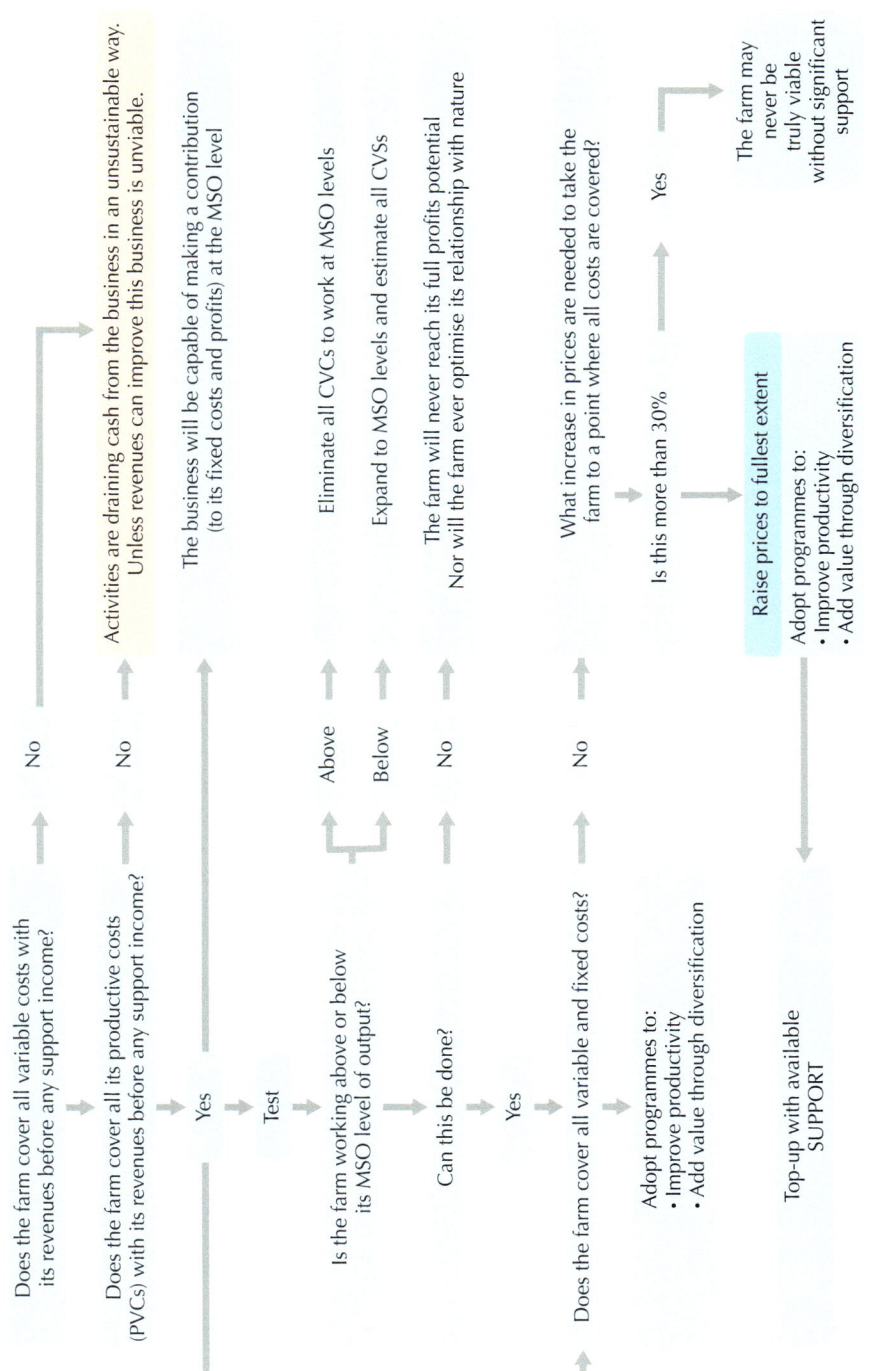

Figure 11.1 The hierarchy of profitability.

When a business can meet its fourth objective it will have access to funds for further investment to expand or to improve its productivity. Each investment scheme should pass the test of paying back within, say, 4 years at the most and delivering a ROTA of better than 15% on the new capital expenditure. Paying back in this context treats the income stream as being, strictly, only the additional revenue benefits (or accredited savings) delivered by the investment.

11.2 SOME WINNING *FORMULAE* FOR SUCCESS

Farm business profitability will be maximised when it works at its MSO point. This requires:

- the elimination of all CVCs. If a farm was above its MSO point beforehand this will take it to MSO
- plus a stocking rate for livestock that uses all the grass available (at around 1.80 LSUs per hectare) or a yield for arable farms that uses all the available nitrogen in the soil.

From an empirical analysis of the Nethergill Database the best-performing farm businesses exhibit the following characteristics, which are also useful targets:

- fixed costs are less than 40% of pre-support income
- PVCs are in the range of 5–15% of pre-support income
- CVCs are in the range of 0–5% of pre-support income.

Fixed costs levels are often determined by scale factors. Many farms are just too small to break below the 40% threshold for fixed costs. In the long term, these businesses must either grow (ideally through aggregation involving contiguous properties) or exit if support payments are inadequate.

When a farm business converges on its MSO position its PVCs seem to reduce to the 5% end of the target range more frequently than the 15% end.

The better-performing livestock businesses exhibit the following common characteristics.

- Grass production is maximised without the use of fertilisers. Unless 8,500 kg dry matter production per hectare is achievable there will be an important competitive disadvantage for commodity-status meat production.
- Livestock are all grass fed.
- The selection of livestock breeds is based on its ability to live outside on a year-round basis, with some winter shelter available for animals to use at their discretion, if necessary, and with adequate shade available on the same basis in summer. In effect, an appropriate breed is one that will consume no CVCs.
- As grass production is the primary driver of output it is important to use all the grass available effectively. This is best done with a mix of cattle and sheep. The most profitable livestock farms are those with a cattle : sheep ratio of 60 : 40 in LSUs.

Some livestock now comes with a **genetic deficit**. For example, some sheep breeds have been developed to produce an increasing number of twin births. While this has obvious benefits in terms of animal numbers if the ewes have not evolved commensurately to provide the extra milk the additional offspring will have to be fed a food concentrate. This concentrate is a CVC. Also, for example, some non-native breeds of cattle are incapable of adequate growth on an all-grass diet. Again, the concentrates are CVCs and, invariably, the premium that can come from either the additional liveweight of the breeds or its brand-value fails to cover the extra costs of the concentrates and the need for winter sheltering.

The best-performing dairy farms businesses exhibit the following common characteristics.

- An optimised balance between the output of milk per cow and its commercial number of lactation cycles.
- Livestock are all grass fed.
- Calves are liquid milk-fed in early months.

- There is a clear separation between the farm proper and the milking parlour in business terms with a realistic transfer price mechanism between the two.
- Slurry production is minimised as a practice rather than accepted and treated industrially.

As the milk output achieved from cows increases, from a no-concentrates base of 4,500 l pa to 12,000 l pa or more, so the consumption of concentrates increases geometrically. Also, as milk outputs increase so the number of lactation cycles reduces from around 7.50 to 2.50. If the lifetime revenue from the cow is to be maximised the best compromise would seem to be 6,500 l pa milk output together with 4.50 to 6.50 lactation cycles. At this level of output some CVCs in the form of concentrates are very likely.

Some dairy farm businesses work on a 100% silage basis and do so quite cost effectively. Apart from the extra cost of silage production, which is a CVC, the main disadvantage is in the slurry management issues becoming a bigger problem.

Some businesses with cheese making facilities can and do work at the low-output end of the spectrum with no concentrate usage. These businesses, by recognising a realistic transfer price and by recovering any extra burdens this might have through premium pricing its cheese products, are suitably profitable. Part of the premium price justification is often the claim to be all grass fed.

Most dairy farm businesses have a contract with a milk supply company. These contracts connect prices to volume guarantees and measures of milk quality. The volume aspect of these contracts is often a barrier to dairy farms moving towards a *lower output-higher lactations cycle* type of regime. All the major milk supply companies have now adopted carbon offset models and will seek to make these available to dairy farm businesses. All these models are different although all will reflect an objective to minimise its collections while growing its outputs. This signals a preference for dairy farms with high output volumes and many farmers have responded to this by pressing outputs with extra concentrates. However, the CVCs effect is to make these farms less profitable as they expand.

Arable farming is driven, essentially, by factors that influence the nitrogen balances in the soil. Most nitrogen enters the soil through the chemistry of death and decay. Nitrides and nitrites (which are compounds with a no or a low oxygen content) form quickly but plants require nitrogen as nitrates (compounds with a higher oxygen content) to be productive and can convert the nitrides and nitrites in due course. This is the process of mineralisation. These conversions take a little more time and effort from Nature. Fertilisers can offer an instant injection of nitrates and make up for any deficiencies in the natural replacement of nitrates in the soil. Excess fertilisers (that is amounts above the maximum available naturally) will increase outputs but only up to a point. Thereafter, evermore fertiliser tends to be needed to maintain this output.

The better-performing arable farm businesses exhibit the following common characteristics.

- Natural mineralisation produces at least 210 kg N per hectare. This should sustain 6.50 tonnes winter wheat per hectare.
- Nitrogen replacement regimes that add another 65 kg N per hectare to sustain a yield of 8.50 tonnes winter wheat per hectare.
- Nitrogen replacement regimes that embrace either a crop rotation element or a livestock element.
- A reversion to non-proprietary/natural seeds that are not engineered to demand proprietary fertilisers.

Many arable farms have removed their former fencing. This is a major barrier to the re-introduction of livestock as a way of replenishing the N balance. Other arable farms, on the urban fringes, will not be able to contemplate any livestock content.

Borrowing for expansion or development is a vital part of any vigorous business. In the farming sector, the prospective borrowers often have the ultimate collateral for banks in their ownership of land. Consequently, there is a common tendency for farm businesses to be over-stretched even for very sound schemes. A measure of the free-cash flow in a farm business should be taken at the level of the 2nd contribution. That is, pre-support revenues *less* variable and fixed costs. Bank managers will use a cover ratio of 4 to establish what level of debt finance to offer. This means that the free-cash flow (at the 2nd contribution) is divided by 4 and the resulting number is considered to be the maximum amount of interest a business can sustain on any debt finance. This will quantify the loan offer.

Long-term loans must be restricted to schemes for expansion or productivity improvement only and the schemes must have an excellent prospect for delivering a good return. An inadequate return will relegate an investment to mere spending. Working capital shortages are the province of short-term loans although these may be renewed on an *evergreen* basis. Borrowings should not be contemplated to correct decapitalisations; that is the role of equity if it makes sense. When cover ratios are below 2 this should act as an alarm signal. Pragmatically, while support payment regimes continue, farming businesses have some in-built insurance as this source of liquidity can be used to cover periods when the free-cash flow is under pressure.

Scale plays an important part in the ability to recover fixed costs and consequently levels of profitability.

To illustrate this, consider two different farms:

- Farm A: 200 ha, revenue generation £500 per hectare, £60,000 fixed costs.
- Farm B: 2,000 ha, revenue generation £500 per hectare, £300,000 fixed costs.

The fixed cost burden of farm A is 60% of revenues; the fixed cost burden of farm B is 30% of revenues despite being greater by a factor of 5.

A farm of 2,000 ha would appear to be a comfortable minimum size to allow sufficient labour to be engaged, adequate equipment to be worked, and all four business objectives to be met.

11.3 DOWNSTREAM ACTIVITIES

All businesses become more profitable when the growth of added-value activities exceeds the attendant added-costs. Dairy farming is a good example of this in action and cheese making takes this one stage further.

Meat processing as a downstream activity is an obvious candidate for livestock farmers to consider. This can range from a ready meals offering more sophisticated meat products or even charcuterie. The immediate requirements for such a business would be:

- a suitable factory
- capability in a credible process
- capital to fund work in progress and stocks
- storage space
- qualified staff.

The immediate operational challenges are to sell the output of the factory remembering that the economics of factories are driven by utilisation factors. Many factories today work on a single 8 hour shift basis for 5 days per week. In contrast, a capital-intensive process industry will work on a 3-shift basis continuously with just 2 of the 21 shifts per week set aside for maintenance activities. All factories will tend to break-even only on high levels of utilisation (at, say, levels in excess of 65%). This output *has to be sold, effectively, on a real-time basis*. Too many excursions into meat processing activities by farmers have foundered on a tendency to store-up all the unpopular or unwanted cuts of meat for a future sale, which then fails to materialise. All the outputs from factories need to be *marketed*.

11.4 MARKETING

There are five classic components to marketing, which is essentially a process of *strategic positioning* to achieve:

- differentiation (in product offerings)
- competitive advantage (through process capabilities)
- grip (on the added-value chain)

- reach (into different market sectors)
- effective structures (for risk-mitigation and management control).

A premium price will only, typically, be achievable if a product has some *unique sales proposition* (USP), which is valued by some part of the market. This is the essence of **differentiation**. The USP can be based on a:

- unique recipe
- distinctive farming regime, or
- sourcing from a distinctive region or landscape.

A guarantee of provenance is fundamental to differentiation. This guarantee cannot be credibly outsourced to another party and must be worked on. However, some differentiation strategies can hide a fatal flaw unless care is taken. This happens when the added-costs of differentiating features are greater than the added-value prospects available in the market. This is the *gold-plating* trap.

Differentiation is re-enforced by product **branding**, which is a mark of authenticity, a signal of quality and an image for consumers. Sometimes brands emerge as new businesses grow and these are then consolidated into a more formal awareness when linked to a commercial proposition. This is where advertising comes in.

Advertising for branded goods is critical but it need not be an expensive proposition. Its objective is to position a particular product offering such that, when a prospective consumer is ready to purchase, the branded product in question is the first to come to mind. Positioning products requires an understanding of consumer behaviour. Broadly speaking, positioning falls into three categories.

- Direct targeting of *obvious* consumer groups with a compelling offer ('shooting').
- Searching for *prospective* consumer groups and making persuasive cases for purchasing ('hunting').
- Attracting the *impulse-buyer* with an easy and unpressurised purchase offer ('fishing').

Most small businesses rely on fishing strategies but while these can be inexpensive, they tend to have low conversion rates. In contrast, very large food producers comprising the category of fast-moving consumer goods suppliers (FMCGs) rely on direct targeting ('shooting' programmes), which are invariably expensive, to pull in sales from its brand enthusiasts.

When a product becomes successful it will attract the attention of competitors wishing to win a share of a new market. Protection comes in the form of **competitive advantage**. This is a way of building **barriers to entry** and can take one of three general forms.

- A patented process, to deny access to competitors.
- Scale of activities, to minimise the unit costs of production.
- Market share, to reduce the costs of production and supply through a phenomenon known as the *experience effect*.

The experience effect is present in all factory situations. In mass-production manufacturing industries it even has a statistical basis. In these cases, it has been found that when cumulative output volumes are doubled unit costs will fall by 23%! This is the driving force behind the growth of market share as a primary commercial objective as this alone will drive the experience effect most rapidly.

With a growing competitive advantage, in the form of lower unit costs of production, a business can either maintain prices and grow margins or discount prices as margins grow and win a greater market share. As always there will be limiting factors. Growing margins will attract more competition and the likelihood of other innovative products emerging. Growing market share inexorably will eventually result in a dominance that can quickly become a vulnerability if market sentiments start to shift.

For meat products, the supply chain from the farm gate to the supermarket-shelf will comprise four main elements:

- slaughter
- butchery
- cold storage
- transport.

The extent to which a farm business participates directly in this supply chain is a matter of **grip**. For many farm businesses wishing to expand into added-value activities the biggest obstacle is the first link in the chain – the abattoir.

Most abattoirs are meat-traders. Animals are purchased on a liveweight or a deadweight basis. Prices will reflect more the demand–supply balance than any aspects of quality or differentiation. Some farm businesses now have their animals killed for a fee as a way around this. Unfortunately, this halfway house solution cannot guarantee an uncompromised provenance. This will only come when mobile abattoirs slaughter on farm properties or when like-minded farmers invest in a dedicated and exclusive abattoir facility.

The next obstacle is butchery. Butchery is a skilled trade and good butchery is always expensive. The real challenge for an aspiring farm-based meat business is to keep a butcher fully occupied (see Section 10.7).

Cold storage is an essential component in the process of matching the patterns of supply with the patterns of demand throughout the year. Again, this is expensive in an obvious sense: building costs and energy consumption charges loom large, as will be the continuous maintenance needed for such a sophisticated asset. Storage is another area with hidden traps. It is all too easy to 'lose' the slow-moving cuts in a warehouse and unless care is taken a store will simply fill up with all the products that are being declined by the market rather than the products being demanded by the market. Residency times need to be monitored carefully and strategies need to be put in place to produce default products, such as mince or sausages, that can return a profit under any circumstances. With such a policy the best cuts will sometimes end up as default products when sales are slow but this makes perfect commercial sense.

The final link in the chain is transport from the cold store to a distribution centre, in the case of large supermarkets, or to a retail outlet directly. Direct deliveries for many small businesses will be essentially a single set of local delivery routes from the cold store base. Local deliveries take place typically within a 40 miles radius from a base. The primary objective of a local delivery operation is to maximise the number of deliveries, in the form of drops, on a single route; the secondary objective is to provide the vehicle and driver with a full days-work. Local delivery routes will encompass any-thing from 6 to 14 drops. The 40-mile limit is the result of trading off the time spent driving to a delivery area against the time spent at outlets making the deliveries. Transport arrangements, especially in small businesses are vulnerable to drivers being unavailable at short notice. In these cases, it is worth recognising that the business may have the ultimate alternative ready at hand – there is nothing more powerful as a statement than to deliver to a customer from the boot of the owner's Jaguar!

A successful food business with a single base and set of local delivery routes will soon yearn to expand and maybe 'go national'. This is the process of **reach**.

Extending the reach of a business requires a plan to exploit a variety of dif-ferent distribution channels. The main distribution channels within the food business are:

- farm gate, farm shop sales
- direct sales to a supermarket chain
- sales to wholesalers
- direct deliveries to retain accounts
- mail-order sales.

Each of these channels is very different in terms of logistics, pricing and fulfilment issues.

Farm gate and farm shop sales will rarely offer outstanding returns but they can be important in marketing terms, useful in product development terms and sometimes just good fun.

Direct sales to a supermarket chain can sometimes be remarkably quick and easy to establish for good products. The downsides are that there will be protocols that may irritate or even offend. For example, some high end stores want their beef with nice white fat and this can only be achieved by feeding cereals to the animals. This puts the committed *all-grass-fed* produc-ers in a dilemma. Other downsides will come when supermarkets ask for greater and greater discounts for the privileges of maintaining volumes and this can be an existential issue if the supermarket has become the single or dominant customer.

Sales to wholesalers will take time to develop as limited amounts will be taken initially as they assess the market. This distribution channel can work well but prices will be keenly set and sometimes settlements can quickly extend to 60 days or more.

Direct deliveries often offer the best margins, but the work-content is prodigious. Running delivery routes is a business in itself. Expansion of this type of business to a national or regional scale can take a very long time.

Mail order has become more common with the growth of online sales platforms. The opportunities opened up by this have still yet to be defined fully.

The activities surrounding the processing of orders and despatching them is now broadly referred to as fulfilment. When things go without a hitch the arrangements seem trivial. However, the test of a good fulfilment system is how it works when corrective action needs to be taken. The first challenge is to handle returns; the second challenge is to chase payment failures; and the third challenge is the greatest of all – organising a product recall when hygiene issues intervene. Getting the responses wrong will be expensive, possibly disastrous.

11.5 ORGANISATION

The role of organisation is to provide structures that will deliver the objectives of the business. There are two different forms to consider. A financial structure that bears the commercial risks of an enterprise and a management structure that uses a chain of command to deploy the physical resources of the enterprise to meet its customers' needs most cost effectively.

Small farm businesses with plans for expansion into new added-value activities will face new challenges in designing and operating new financial and management arrangements.

Most farm business units would be classified as being small (even very small) in the context of other business enterprises. Farmers have sometimes addressed this by entering into various forms of cooperative schemes. Examples can be found where small groups of farm businesses have arrangements to share plant and equipment. Other examples can be found where farm businesses join cooperatives to market farm produce. These arrangements have a track record of mixed success. Produce cooperatives often reflect the quality values of the weakest members – the lowest common denominator effect – and disappoint when it comes to profit sharing. A better model is likely to be based on working under a collaborative approach (implying the need for a powerful and well-administered brand) where a charter sets out all the essential quality protocols.

Small businesses have the advantages of *executive-owners* and short lines of command. These are *power-centred* organisations and the best are very good indeed. When businesses grow to the point of requiring salaried executives the executive-owner model will have evolved into a *merit-centred* organisation and the original executive-owners often feel they have been sidelined. At this point it will be important to remember that the participating farm business units are the ultimate guarantors of quality and provenance.

As small farm businesses expand their commercial horizons the immediate organisational challenges will centre on fulfilment capabilities. This is the sharp end of marketing.

EPILOGUE

Despite all the challenges facing the farming sector its prospects must be rated as very good. There will always be a demand for food. This might seem so obvious that it requires no comment, but other industries come and go as the economy changes.

The UK, with the Republic of Ireland, is possibly the best grass producer in the world. It has a managed landscape that has evolved over centuries to fulfil this role superbly. The economy can do no harm in maximising the availability of commercially competitive grass for conversion into meat products by ruminants. This is not to crowd out other types of farming. Cereals have a place, too, but **for human not animal consumption**; and woodland has a role to play especially on the margins of productive land.

Grass production must come first, and the economy cannot afford to neglect a single hectare of pasture nor divert good pasture to alternative uses, with exceptions for essential infrastructure or issues such as flood mitigation. In this regard, **essential infrastructure should not include renewable-energy schemes**. Farming is the leading component of the energy sector in the economy and a ruminant on grass is a more efficient converter of energy into food than any electricity supply scheme can match.

Climatic conditions worldwide are always in a state of flux. Despite political and social sentiments to the contrary the UK will experience a warmer and wetter climate in the coming years. Biodiversity will explode without any human intervention and farming will have to change to accommodate what should be regarded as an opportunity. Already, some experts predict that the best sparkling wine in the world will in future come from the vineyards of Kent, Sussex and Surrey.

It must be expected that the system of farm subsidies will end; this will come sooner rather than later. Today, in 2024, taxation is at a generational high point and public spending is under great pressure. Subsidies in any form will distort the marketplace and the economy in general. Small farms that are incapable of covering variable costs will disappear. This will be an opportunity for other farms to expand through a process of aggregation. Other farms that fail to cover both variable and fixed costs will slowly decapitalise out of existence. This can only be good for the economy in the long run.

However, this is not all gloom. **The removal of subsidies will promote a rise in farm gate prices, and this will alleviate and transform many farm businesses**. After an initial public outcry about food prices the issue should quietly disappear. A 50% increase in farm gate prices might only result in a 5% increase on the shelf if the other components of the food supply chain maintain its own particular added-cost structures.

Food security is an issue of national strategic importance. Output volumes alone is not the answer to this question. It is much simpler; **the best guarantor of food security is farm profitability**. Perhaps surprisingly, if profitability is maximised in the farming sector the ultimate volumes will be maximised, too.

APPENDICES

A.1 STANDARD DIVISIONS OF THE UK

The landscape of the United Kingdom can be classified into a number of different zones (see 10.16) to reflect the prevailing and distinctive features that comprise each district. One such classification was set out by G.H. Drury in his seminal work on the geography of the British Isles. A modification of this scheme by the authors, to comprise 32 divisions for the UK, is shown in Figure A.1.

For each of the divisions there is a reasonable expectation that a scheme for land classification will deliver a consistent set of quantifiable values. For example, 'unimproved pasture' will produce much the same in *kilograms of dry-matter per hectare* across the division. Therefore, it is possible to consider defining a series of standard farms for each division. These farms should comprise, at the very least:

- three livestock businesses
    - all sheep
    - all cattle
    - mixed livestock
- a dairy business
- two arable businesses
    - all arable, predominantly wheat
    - mixed arable and livestock, predominantly wheat.

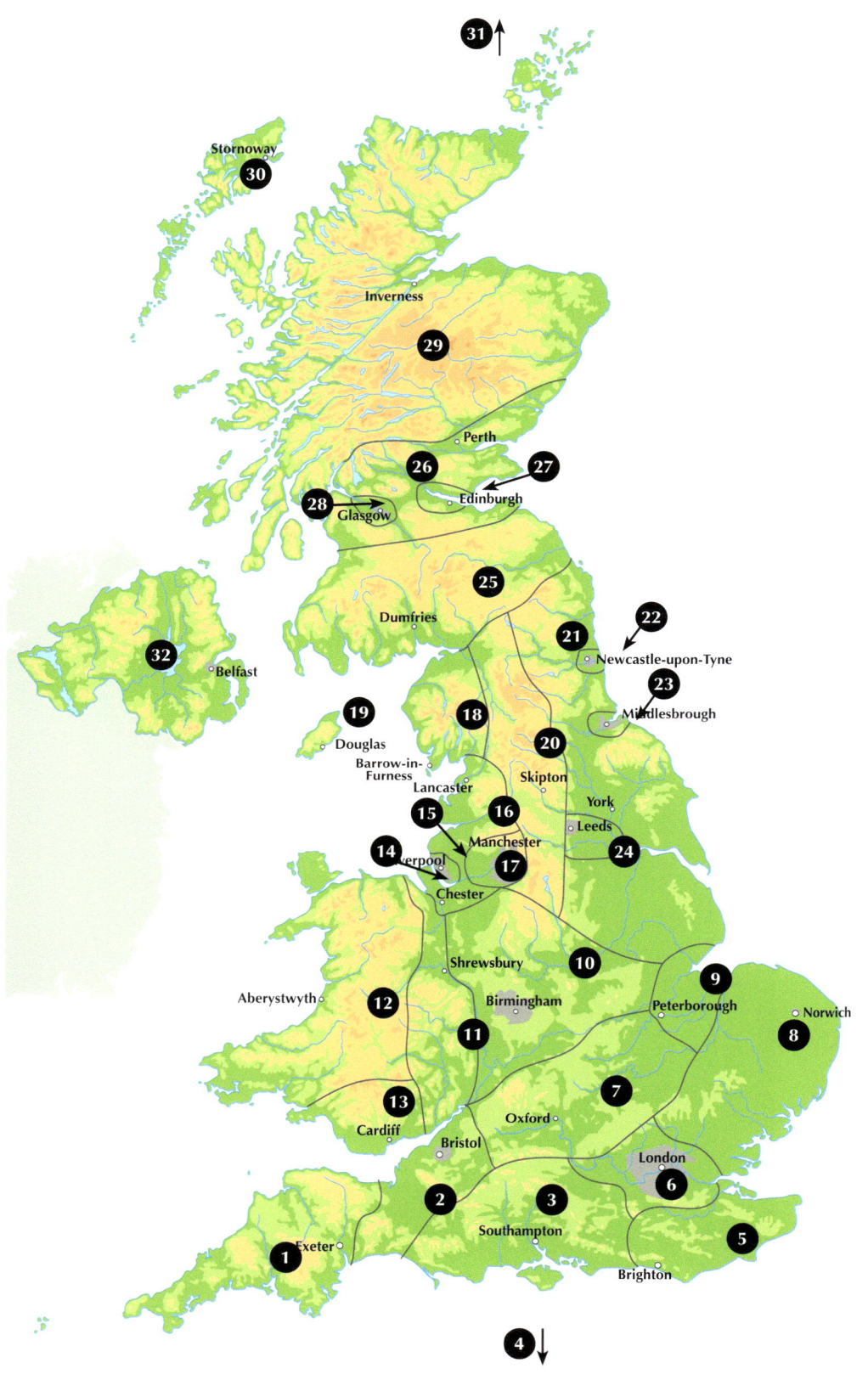

Figure A.1 Map of natural regions of the UK.
After: G.H. Drury.

1.	The South-West of England	Exeter
2.	Wessex	Bristol
3.	The Solent Basin	Southampton
4.	The Channel Isles	St Helier
5.	The South-East of England	Brighton
6.	The Thames Basin	London
7.	The Scarplands of the South Midlands	Oxford
8.	East Anglia	Norwich
9.	The Fenlands	Peterborough
10.	The Midland Triangle	Birmingham
11.	The Welsh Marches	Shrewsbury
12.	The Welsh Massif	Aberystwyth
13.	The South Wales Coalfield	Cardiff
14.	The Cheshire Plain	Chester
15.	Industrial Mersey-side	Liverpool
16.	The North-West of England	Lancaster
17.	Industrial Lancashire	Manchester
18.	The Lake District	Barrow-in Furness
19.	The Isle of Man	Douglas
20.	The Pennines	Skipton
21.	The Scarplands of Eastern England	York
22.	Industrial Tyne-side	Newcastle-upon-Tyne
23.	Industrial Tees-side	Middlesbrough
24.	Industrial West Riding	Leeds
25.	The Scottish Borders	Dumfries
26.	The Lowlands	Perth
27.	Metropolitan Edinburgh	Edinburgh
28.	Industrial Clyde-side	Glasgow
29.	The Highlands	Inverness
30.	The Western Isles	Stornoway
31.	The Northern Isles	Lerwick
32.	Northern Ireland	Belfast

As farm sizes vary considerably throughout the country, each farm type in each division should reflect the prevailing average.

Model farm accounts for each standard farm should be cast at its MSO output level and on a tenancy basis, reflecting a fair rent. These models would serve as benchmarks for performance targets for individual farm businesses.

The work involved in setting up such a scheme is prodigious but small in comparison to the great weight of farming sector data already collected in different forms. It would be ideal for some university to adopt as a basis of a research programme.

A.2 ROTA: THE TELEVISION-TUNING ANALOGY

The ROTA concept is a significant challenge for many people in business (other than farmers). Money is not generally perceived to be a *two-dimensional* (2D) phenomenon with aspects of amount and term. The two dimensions of ROTA (see Section 3.2) are **assets-turn** and **margins.**

Being 2D allows ROTA to be plotted on a chart. Each plot will be a distinctive measure of performance and the question then arises as to how might performances be improved? The answer will always be 'move as far to the northwest (top right-hand corner) as possible'.

This can be done by tackling items on the B/S, first, to improve assets-turn and then the P&L to improve margins. To do this, some items will need to increase and others to decrease. The process is analogous to the former problems of tuning a television to produce a perfect picture. This involved tuning the vertical-hold and then the horizontal-hold. The vertical-hold on the ROTA chart is the assets-turn and the horizontal-hold is the margin.

The 'knobs', which represent items on the B/S and P&L, can now be turned in the indicated directions for performance improvement. The B/S knobs will move the plot vertically and the P&L knobs will move the plot horizontally. The combined movement will move the plot diagonally to a new performance level.

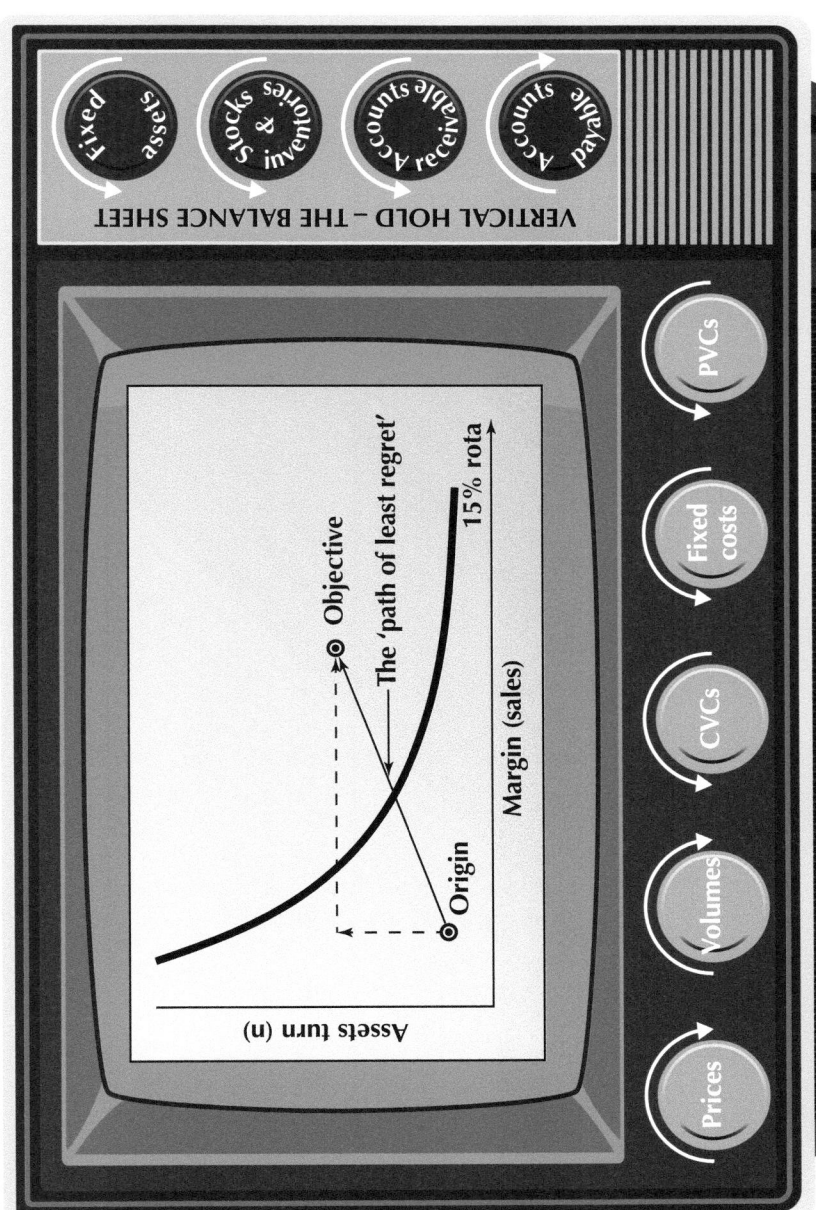

Figure A.2 ROTA: television tuning analogy.

A.3 NET PRESENT VALUES (NPV)

All income streams over time will have an equivalent capital value based on views taken about interest rates, business risks and the time horizon for the income. This capital value is defined as the net present value of its cash flows discounted over the time horizon.

What is discounting? Very simply, if interest rates are 5%, the value, today, of £100 next year is just £95.00. The formula is, if x is the interest rate:

Value (next year) = £100 / (1 + x / 100)

The year after, the £95.00 will be worth £90.25 and so on.

What is net present value? If the income stream of £100 was available for just 3 years, its net present value will be:

NPV = £100 + £95.00 + £90.25 = £285.25

That is, if a person paid £285.25 for three years of income at £100 they would come out even.

What is horizon and its significance? Horizon is simply the number of years the income stream will be maintained.

Suppose the £100 was the income from a business investment. Instead of applying the interest rate as the discount a business should look for 15% to compensate for the risks involved. A £100 income stream at a 15% discount over 15 years has an NPV of £608.43. If the horizon was extended to 20 years, the NPV would be £640.83. Further, if it was decided that 20 years might be feasible but only 15 is realistic the **residual value** (RV) at 15 years is £32.40 (being the difference in the NPVs). This is only 5.32%.

NPV calculations are useful for evaluating investment projects. The income stream at any time is the net value of the revenues and the expenditures. In a project, the early years are heavy on expenses and light on revenues resulting in a negative income stream. This must be offset by later income streams, which need to be positive. If a project horizon is, say, 15 years there will be a discount rate that delivers less than 1% residual value at this point. This discount rate is termed the **internal rate of return** (IRR) on the project. The year at which the NPV becomes positive is known as the **pay-back period**. A business might have an investment policy which, for example, might specify that all projects must pay back within 4 years and have an IRR of at least 15% over a 15 years horizon.

A.4 PE RATIOS

As shown earlier, a £100 income for 15 years is worth £608.43 in today's values with interest rates at 5%. The ratio £608.43 / £100, which is 6.08 is known as the PE ratio. The term comes from the world of finance and represents a price-to-earnings (P/E) ratio. It is useful to convert earning into capital equivalents, or vice versa, very quickly.

In this book, when farm income streams have been capitalised a PE ratio of 4.00 has been used. This is equivalent to expecting a 15% ROTA for an horizon between 5 years (PE = 3.71) and 6 years (PE = 4.15). This is a fairly short horizon and the horizon is often taken to measure the **quality of earnings**. Construction contracts are taken to deliver a poor quality of income for a contractor and many have PE ratios of just 1.50. An electricity supply contract for 25 years, however, would rate a PE of 6.50, so 4.00 is a reasonable expectation for a conservative business. Speculation on the stock market often results in very high PEs (over 20 for hot digital stocks) but in these case speculators are valuing stocks on an expected growth in earnings in the short term.

A.5 QUALITATIVE VERSUS QUANTITATIVE MEASURES

Measuring biodiversity is fraught with difficulties and biodiversity net gain (BNG) is even more challenging. Almost all measures of biodiversity are qualitative. That is, specifications are essentially descriptive and, whilst these are meaningful to experts, the measures fail when comparisons need to be made. This is the problem of answering *how many worms equal a lapwing?* A halfway-house is simply to count species but this implies that the greater the number of species the better the environment. In a managed landscape there will be an optimised (uncompromised) pattern of biodiversity and increasing the species numbers in this case would throw the landscape out of equilibrium. When this happens, a dominant species will emerge and prevail until a new equilibrium is found. That is, beyond a certain point in a managed landscape an increase in biodiversity can prompt a later fall in numbers.

A necessary condition for all quantitative measures is that a common unit of measure is used. Such a measure for the environment is not obvious. However, despite the sentiments of many people with concerns about the quality of the environment, ultimately its value must be essentially one of economics. The landscape produces an income stream for the economy and this should be the basis of its value. If the income stream is valued at the MSO point for all farms it can be said that maximising profits is consistent with optimising biodiversity and that the resulting pattern will not be compromised in any way (see Chapter 6).

A.6 VARIABLE COSTS THAT ARE DIFFICULT TO CLASSIFY

Variable costs are **consumables.** They can be either labour-related or supply-related (inputs and purchases). MSO theory requires these variable costs to be split into productive variable costs (PVCs) and corrective variable costs (CVCs).

PVCs are typically those involved in working with Nature, and CVCs are those involved in substituting for Nature together with any costs over and above the minimum necessary as a consequence of natural disadvantages or weather events. This division is analogous to the practices of industry where costs incurred over and above the standard, cast on a zero-based-budgeting basis, is a cost variance and *physical noise*. In this context, *noise* is any work over and above the minimum necessary to complete the task. In farming, many CVCs are characterised by the industrial energy content of their manufacture.

A table of variable costs that are difficult to classify together with the classifications used in this book and the reasoning applied is shown in Figure A.6.

Figure A.6 Guidance on cost classifications.

Variable costs	Classification	Notes
Bedding	CVC	Farm-regime-related and optional
Cereal-based feeds	CVC	Farm-regime-related and optional
Concentrates	CVC	Contains industrial energy
Elective Vet & Med	CVC	Farm-regime-related and optional
Emergency Vet & Med	CVC	Event related
Essential Vet & Med	PVC	Essential for uncompromised animal welfare
Fertilisers	CVC	Contains industrial energy
Irrigation	Balance sheet	Alters the physical asset and is a capital issue
Lime	Balance sheet	Maintains the physical asset in a managed landscape
Mob grazing	CVC	Farm-regime-related and optional
Seeds (arable): purchased	CVC	A substitute for natural seed retentions
Seeds (arable): retained	PVC	Necessary minimum
Seeds (grass)	CVC	A substitute for natural grass growth
Walls & fences	Balance sheet	Alters the physical asset and is a capital issue

NOTES

1   *Killing the Countryside*, Graham Harvey, Vintage, 1998.
2   https://www.bcg.com/publications/2023/regenerative-agriculture-benefits-germany-beyond.
3   https://www.whatthesciencesays.org/are-there-only-100-harvests-left-in-british-soils/.
4   The sun delivers to the Earth 1.36kW of energy per square metre. This is the total solar irridance (previously refered to as the solar constant). (https://earthobservatory.nasa.gov/features/EnergyBalance). Solar energy has three main components:
    - The visible light spectrum, which is invariant and drives photosynthesis.
    - The infra-red spectrum, which decreases as a proportion with latitude and drives the heating effect.
    - The ultra-violet spectrum, which increases in proportion with latitude and is associated mostly with harmful effects.
    The intensity of solar radiation received will change with latitude and it is quoted that the UK receives on average 101.2 W per square metre, within a range of 71.8 to 128.4 W/m2 (Dougal Burnett, Edward Barbour, Gareth P. Harrison, 'The UK solar energy resource and the impact of climate change', *Renewable Energy*, 71 (2014), 333–343, https://doi.org/10.1016/j.renene.2014.05.034). This is effectively the heating effect which is quite separate from photosynthesis.
5   Ammonia Technology Roadmap, IEA, https://www.iea.org/reports/ammonia-technology-roadmap
6   https://sustainablefoodtrust.org/news-views/the-hidden-cost-of-uk-food-soil-degradation/
7   https://www.nethergillassociates.co.uk/Portals/0/Farming%20at%20the%20Sweet%20Spot_1_1.pdf; https://www.3keel.com/wp-content/uploads/2022/08/future_of_feed_full_report.pdf.
8   https://www.nethergillassociates.co.uk/Portals/0/Farming%20at%20the%20Sweet%20Spot_1_1.pdf; Defra, Agriculture in the UK 2020 (2021), https://assets.publishing.service.gov.uk/media/6331b071e90e0711d5d595df/AUK_Evidence_Pack_2021_Sept22.pdf.

9   https://pdf.euro.savills.co.uk/uk/commercial-retail-uk/spotlight-uk-grocery---january-2022.pdf Global data; https://www.gov.uk/government/statistics/agriculture-in-the-united-kingdom-2022/summary.

10 https://ourworldindata.org/grapher/annual-co2-emissions-per-country?country=~GBR.

11 University Corporation for Atmospheric Research: https://web.archive.org/web/20210604114815/https://scied.ucar.edu/interactive/earths-energy-balance.

12 University Corporation for Atmospheric Research: https://web.archive.org/web/20210604114815/https://scied.ucar.edu/interactive/earths-energy-balance.

GLOSSARY

Albedo

A significant amount of sunlight is reflected by the different surfaces that comprise our world. Only perfect 'black-bodies' absorb all incident energy without reflection. The albedo factor is taken to be 0.30 currently. That is, 30% of incident sunlight is reflected without warming the ground. This will change, and increase, if cloud cover increases. It will also change and decrease as the icecaps melt.

Arrow of time

This term is associated with the concept of entropy. All systems, such as Nature, move in time from a particular state of order to a new state of less order. In doing so it is said that the entropy of the system has increased. In becoming less well ordered a system decays and the cause of the decay is often referred to as the arrow of time. It signifies that time only moves in one direction (i.e. it is not reversible) and therefore decay in a system is inevitable.

Assets turn

Money is a two dimensional entity. It has amount and it has term. Amount is associated with the profit and loss account (P&L) whereas term is associated with the balance sheet (B/S). Margins in business are measures of profits as a percentage of sales. However, this will not compare directly with interest rates on deposits, which are a measure of returns on capital or asset values (ROTA). To convert margins into a measure of ROTA a multiplier known as assets turn (AT) is used. AT is simply: sales / assets employed. AT measures the sales volume required to cover the value of the assets employed in a business.

An assets turn of 2× would signify that the assets employed have a value equal to half the sales volume. Another way of expressing this is to say that the assets employed are equivalent to 182.5 days-of-sale equivalent. This illustrates the term aspect of the B/S directly.

Attenuation

If the environment is adulterated by inappropriate farming practices this is equivalent to decapitalisation. The income stream from natural resources is said to be attenuated when this happens. As the drivers of the attenuation process will be the CVCs and fixed costs in a business, profitability will decline as natural capital is effectively being destroyed.

Break back

CVCs, if present in a business, will be incurred after the resources related to PVCs have been exhausted. These CVCs will, almost invariably, increase at a faster rate than revenues. (The reasons for this are discussed under the MSO and Thermodynamics headings later). As CVCs increase there will be a point when total costs overtake total revenues again. This is the break-back point. Beyond this point all the accumulated profits will have been eliminated and losses will increase indefinitely.

Break even

The revenues and costs in a business will increase with output volumes, but in different ways. Revenues increase generally as volumes increase in a linear fashion. Total costs, which comprise fixed and variable costs, also increase with volumes but only as variable costs are incurred. At low volumes of output total costs will exceed total revenues but there will be a point, in a viable business, where revenues will equate with costs. This is the break-even point. Beyond this break-even point revenues will exceed costs to an ever-greater extent, at least up to some future limiting point.

Cluster

A small group of farms, with each residing in near proximity, is often referred to as a cluster when they cooperate or are associated in some way.

Discounting

A sum of money today, say £100 on deposit in a bank, will have a different value next year when interest is added to the account. If interest rates are

5% the value of the deposit next year will be £105. Conversely, an income stream of £100 pa for 5 years will not be worth £500 today. From today's perspective the £100 will only be worth £95 next year (if interest rates are 5%) and only 95% of £95 the year after. This is the process of discounting. The discount rate in such circumstances will be the same as the interest rate. However, if a business has an objective to deliver a return on capital of 15%, this will be the discount rate to apply.

Distress prices

In commodity markets prices are set by balancing demand with supply. These will equate at some price and this is the market price. While prices can move dramatically as both demand and supply change the fluctuations are accommodated in the pricing mechanism. However, suppose a business goes bankrupt and all its stock is to be sold to pay off creditors. If this stock is large in relation to the usual levels of demand the market will not absorb the extra volume without serious disruption. In these cases, stock is often sold with big discounts. This is distress pricing.

Evergreen basis

This is a term to signify ongoing, forever or in perpetuity. In practice it will mean no change over the foreseeable future.

Entropy

When systems decay, it signals a growing state of disorder and this decaying process is linked to the passage of time. This happens both in Nature, where energy levels dissipate as one form of energy, such as sunlight, is transformed into another form, say heat, and in mechanical systems where, say heat is used to produce steam for mechanical power. Scientists measure the degree of disorder in a system by its entropy as disorder grows so the entropy value increases. It has no conventional units of measure and only differences tend to be quantified. No absolute measures have emerged and entered common use as the task is essentially impossible; entropy as an absolute value cannot be separated from a measure for the universe as a whole.

Why is such an academic concept important to farming? The general answer to this is that it provides a basis for identifying impossible, or mutually exclusive, objectives. For example, the industrial energy content of some supplies, such as fertilisers, can never be recovered in the food that eventually gets produced. The use of fertilisers have been justified by economists on cost grounds not energy considerations. As food is a low-value commodity and electricity, for example, is a high price consumable, the economics soon run out. Most situations like this which are justified economically are based on the use of products which are by-products of some other process.

It might also be asked, that if the entropy phenomenon of decay and disorder prevails, how do societies create civilisations which are clearly based on the creation of order? Decay does not preclude the creation of order in some small sub-system; all that happens is that the price of this order (in negative entropy) is more than offset by a greater amount of disorder for the residual universe. This we can conveniently ignore!

Farm output
The revenue from the sale of farm produce before any support payments are taken into account.

Fixed costs
The permanent costs of establishment in running a farm business. These costs will not change with different levels of output.

Flying herd
Most livestock farms breed animals. These animals are nurtured until they are finished for slaughter or are available for milking. This allows farmers to control the quality of their herd and eradicate diseases. Some farmers will simply buy in animals simply to finish them quickly or to put them directly into milking. This practice avoids all the risks in breeding and reduces the time before animals can play a commercial role. Such animals will comprise a flying herd. In operating a flying herd, a farmer is pushing some business risks down the supply chain to breeders (and in doing so will be blind to a breeder's practices – which can be heavily dependent on CVCs). Also, a flying herd, as a result of its mixed nature, can deliver animal health downsides and problems with product consistency.

Free issue
Nature, in providing farms with sunlight, rainfall and fertile soils, is delivering valuable resources essentially free, although it will take work to exploit

these advantages (in the form of PVCs). These resources are therefore available on a free-issue basis and will not appear as costs in any set of accounts.

Hook-curve effect
When the characteristics of a population (such as animals, customer accounts or profits) exhibits randomness a corresponding phenomenon will always be present – the hook-curve effect. If the population is ranked in descending order of value and the cumulative value is recorded it will result in a hook-curve. That is, the cumulative value grows quickly to a peak and then slowly decays thereafter. The end point, which is equivalent to the net value of the population, can often be negative.

Inflexion point
The variable costs in farming have been decomposed into two sequential elements in MSO theory. These components are the PVCs and the CVCs. The point at which the PVCs end and the CVCs commence is an inflexion point in the variable cost line. Typically, an inflexion point occurs when the behaviour of an equation changes from one formulation to another.

Lairage
Lairage comprises an area set aside for animals before slaughter. A good lairage will keep the animals clear of any concerns and keep them quiet.

Laissez-faire
The celebrated economist Adam Smith postulated that there was an 'invisible hand mechanism' in a free market that would always mean that the needs of the marketplace would always be satisfied as suppliers respond to demand. The free market system referred to an absence of intervention by the state or a cartel and the policy of non-intervention became known as laissez-faire (French for leave-alone).

Law of diminishing returns
Suppose the price of a product stays constant in a marketplace but the cost of production increases as more output is delivered. This happens, for example, when a limited supply of minerals ores won through mining reaches the end of its life. The easier ore to mine is the first and the last will be more difficult. The profits from this situation will decline steadily as more output is delivered. This is the law of diminishing returns.

Laws of thermodynamics

Farms produce food and foods are a fuel. Fuels are a form of energy and the science of energy is founded in thermodynamics. There are four laws in thermodynamic theory which constrain all aspects of the production and uses of energy. These are set out below, phrased for farming (Rules A,B,C and D) but referencing the illogical numbering used in science (0th, 1st, 2nd, and 3rd).

- **Rule A (0th Law)** All landscapes in contact with Nature will, at some point, reach a position of common equilibrium with Nature. This is the managed landscape.
- **Rule B (1st Law)** To transform the managed landscape from one state to another cannot be done without putting in work. Change does not come free – even a 'do-nothing' form of re-wilding.
- **Rule C (2nd Law)** All change creates disorder and disorder increases with change. These changes will never be fully reversible. Therefore, the energy content of farm inputs, such as concentrates and fertilisers, can never be fully recovered in the resulting farm produce, as food calories.
- **Rule D (3rd Law)** There are no pathways back to any set of original conditions. Change, such as a re-wilding programme, can only produce a new future; it can never return to some previous past.

MSO

MSO is the point of maximum sustainable output. This is the farm output generated when all the natural resources available on the farm have been consumed and no artificial input has been used. For most farms this will be at the inflexion point between the PVCs and the onset of the CVCs. The exceptions to this are those low-output farms where not all the natural resources have been consumed. It is worth noting that a low-output farm may still consume CVCs despite having unused natural resources.

Natural capital

All resources, including natural resources, can be regarded as capital items. As such these resources are regarded as assets in the economy. Many discussions around the concept of natural capital are in the context of offsetting qualitative values against hard-nosed financial capitalism or quantitative values. However, natural capital can be and must be quantified. It can be defined as the NPV of the income stream attributable to the use of natural resources in farming to produce food products. In turn, the relevant income stream for a farm can be defined simply as farm output less PVCs.

Negative feedback

Consider the springing on a car. When a vehicle hits a bump in the road the coil springs will compress to even out the ride. The coil springs are designed to act quickly in response. However, this very feature means that the spring will continue to 'bounce' up and down (oscillations) after the bump creating an unpleasant ride. A shock absorber addresses this problem. A shock absorber is designed to act slowly so that a sharp movement by the coil spring is offset by the resistance of the shock absorber. This dampens out the oscillations and improves the ride. This is an example of negative feedback. In general, some part of the output is fed back into the system to reduce the output. Most natural systems have feedback mechanisms which stop events getting out of control. Global warming will produce more moisture in the atmosphere and this, in turn, will increase cloud cover. An increase in cloud cover will result in a greater degree of reflection for sunlight (see albedo) and this will reduce temperatures (but by an amount less than the increase in the first place).

Net present value

This is the amount that might be paid, in a single instalment, to secure an income stream in perpetuity. As money tomorrow is less valuable than money today it gets discounted by the prevailing rate of interest. The sum of these discounted cash flows is the NPV of the income stream.

PE ratio

This term stands for price-earnings ratio and comes from the world of finance. Consider an income stream of £100 in perpetuity. What might someone be willing to pay, in a lump sum, for the income stream? If it was thought that a fair offer for the income stream is £400, the PE ratio would be 4. PE ratios convert income streams into capital and vice-versa. The actual ratio will depend on the rate at which the income stream has been discounted. A high discount leads to a low PE ratio and vice-versa.

Positive feedback

Consider a bank extending a credit facility to a valued customer. If £100 is offered as a facility the bank is taking the view that it will be paid back according to the terms of the loan. However, the contract that specifies the terms is a security asset and will be added to the asset base of the bank. By lending the money it reduces its liquidity but increases its assets base and is able to consider risking more liquidity on other loans as it has the security to do so. This is a positive feedback mechanism as the output of loans grows as more loans are extended. However, this cannot go on indefinitely as eventually liquidity runs out. When banks fail to recognise this the consequence is bankruptcy.

Positive feedback mechanisms can be found in Nature where there is a pattern of chaotic behaviour. Suppose it is decided to re-wild a managed landscape by just leaving it alone for a number of years. The response by Nature will be chaotic as options to change appear randomly and as Nature chooses (always) to take the path of least resistance. A dominant species will always emerge in these circumstances until a new equilibrium is achieved. The dominant species benefits from an 'unfair/unexpected' advantage while Nature is in transition while other species only experience disadvantage from the change.

Profitability

In general, profits are the surpluses in revenues after costs have been deducted. However, this is far too general to be useful. Profitability can be regarded as a hierarchy of different levels of contribution, each with a distinctive role. See below.

- **1st contribution** Farm output less total fixed costs. It is the contribution available to cover all subsequent costs and distributions such as fixed costs, farm drawings, capital expenditures, finance charges and tax liabilities. When 1st contributions are negative, a business will lose cash on every transaction and will be unviable without subsidy.
- **2nd contribution** 1st contribution less total fixed costs. It is the contribution after the other remaining costs in the business. When this is negative, the business is decapitalising and will only survive if capital injections are made from time to time.
- **3rd contribution** Unique to farming, being: 2nd contribution plus all subsidies, support payments, and grants (net of consequential expenditures). This often comes closest to the profits declared in

statutory accounts, once the provisions and accruals applied by accountants are eliminated.

Random walk

Consider tossing a coin in a game where heads means a right-hand turn and tails a left-hand turn. After, say 50 tosses of the coin, the path taken is termed a random walk. Nature behaves this way. Not unsurprisingly, if a person turned around to face the opposite direction after the 50th coin toss and applied the same rules for the next 50 coin tosses the person has a vanishingly small chance of ending up in their original starting position.

Resonance

Chaotic behaviour will spontaneously produce order at various points in its 'career'. The order is transitory and mostly lost as events continue apace. However, in rare circumstances, other things may intervene, and the order is crystalised into reality. This is the phenomenon of resonance. When geological timescales are applied the results are abundant. Many geological features (comprising the 'wonders of Nature') happen this way.

When a habitat changes to a new and stable form it will be the consequence of resonance resulting from a new equilibrium.

ROTA (return on total assets)

There is an old verse that goes 'money matters are a function four; it is a medium, a measure, a standard and a store'. Money has two aspects (dimensions, in technical terms): it has a value and it has a term. This is encapsulated simply in the concept of interest paid on deposits. When the rate is, say, 5% it is taken that this means 5% for 1 year. Over 2 years the return will be 10% on a simple basis.

Businesses are in a competition for capital and must do better than the deposit rates at a bank. However, it is not obvious how comparisons might be made. ROTA resolves this problem as it equates, conceptually, to an interest rate. ROTA comprises two components; firstly, profit margins on sales (measuring amounts) and, secondly, assets turn (measuring term). Profit margins on sales equate to profits/sales and assets turns are sales/total assets. When multiplied together the result is ROTA which equates to profits/total assets.

Saddle points

In Nature it is not uncommon for some phenomena to move in one direction while others will move in different or even opposite directions. This promotes a trade-off situation and a search for an acceptable compromise between the different phenomena. When farm profits are being maximised it would be important to confirm that this would not be at the expense of the environment, or profligate in energy consumption. If profitability was maximised, with no environmental damage, and energy consumption was minimised the system would be positioned at a, so-called, saddle-point in Nature. This derives its name from the three-dimensional shape that the equations resemble. One arc sits (downwards) over the horse and the other, at right-angles provides (upwards) support for the rider. Where the shape tops-out for the horse and bottoms-out for the rider is the 'mathematical' saddle-point. MSO is the saddle-point where profitability is maximised, Nature is optimised for output, and energy consumption is minimised. Nature is optimised in the sense that it is uncompromised, but it will differ from any other landscape.

Standard model

The standard model of the firm is an economic model that evaluates profitability. It has three essential components: a revenue line which is linked to output volumes; a burden of fixed costs which are independent of output volumes; and a line of variable costs which is also linked to output volumes. In most cases, the revenues and variable costs will be linearly related to output volumes.

Unit costs

Unit costs are simply the total costs of production divided by the volume of output secured. Its usual measure is £/unit of output. However, unit costs will change as cost profiles and output volumes change. On the standard model of the firm the unit cost will decrease, at reducing rates, as outputs grow. On the MSO model, where variable costs are decomposed into PVCs and CVCs, the reductions in unit costs only occur up to the MSO point. Thereafter, unit costs will increase.

Variable costs

The cost inputs in a business that are volume dependent. In MSO theory, these costs decompose into two sequential groups. These are the productive variable costs (PVCs) and the corrective variable costs (CVCs).

- **PVCs (productive variable costs)** Those exclusively associated with working with Nature and the other natural resources on the farm. These costs would include grass management, ploughing and harvesting. PVCs are linked to free-issue resources.
- **CVCs (corrective variable costs)** Those associated with substituting for Nature or other natural resources. Typically, these cost items will be one of two types:
    - items that incorporate an industrial energy content
    - items that correct for some disadvantages linked to latitude, elevation or rainfall (which would include concentrates, fertilisers and silage production).

FURTHER READING & LISTENING

A Sand County Almanac, Aldo Leopold, Oxford University Press, 1949

Adam Smith: What He Thought and Why it Matters, Jesse Norman, Penguin Random House, 2018

Agriculture in the UK: Evidence Pack, Government Statistical Service, Defra, 2022

The Amateur Poacher, Richard Jefferies, Oxford University Press, 1978

Biophysical Economics: From Physiocracy to Ecological, Boston University, 1999

The British Isles, G.H. Drury, Hieneman, 1972

Calculating Planetary Energy Balance & Temperature, Center for Energy and Environmental Studies, https://scied.ucar.edu/earth-system/planetary-energy-balance-temperature-calculate

The Carbon Fields, Graham Harvey, Grass Roots, 2008

Dirt, David R Montgomery, University of California Press, 2008

Economics and Industrial Ecology, Cutler J. Cleveland, Department of Geography and Environment, Boston University, 2007

Einsteins Fridge, Paul Sen, Williams Collins, 2021

The Energy Collapse: Louis Arnoux, Podcast: Rachel Donald, Planet Critical, 2024, https://www.planetcritical.com/p/the-energy-collapse-louis-arnoux?utm_campaign=post&utm_medium=web

English Pastoral: An Inheritance, James Rebanks, Pengui, 2021

Entangled Life, Merlin Sheldrake, Penguin Random House, 2023

Fantastic Fungi, Paul Stamets, Earth Aware, 2023

Feral, George Monbiot, Penguin Books, 2014

The Feynman Lectures on Physics, Feynman, Leighton, Sands, Caltech's Division of Physics, Mathematics and Astronomy, 1961,https://www. feynmanlectures.caltech.edu/I_03.html

The Future of Feed: How Low Opportunity Cost Livestock Feed Could Support A More Regenerative UK Food System, Julian Cottee, Caitlin McCormack, Ella Hearne, Richard Sheane, WWF, https://www.3keel.com/wp-content/uploads/2022/08/ future_of_feed_full_report.pdf

The Gamekeeper at Home, Richard Jefferies, Oxford University Press, 1978

The Glorious Life of the Oak, John Lewis-Stempel, TransWorld Publishers, 2018

Green and Prosperous Land, Dieter Helm, Cambridge University Press, 2024

The Hidden Cost of UK Food: Soil Degradation, Megan Perry, Sustainable Food Trust, 2024, https://sustainablefoodtrust.org/news-views/the-hidden-cost-of-uk-food-soil-degradation/

Holistic Management, Allan Savory, Island Press, 2016

The Illustrated History of the Countryside, Oliver Rackham, George Weindfield & Nicholson Ltd, 1994

Key Management Ratios, Ciaran Walsh, Pitman, 1996

Legacy, Dieter Helm, Harper Collins, 2020

The Lost Rainforests of Britain, Guy Shrubsole, William Collins, 2022

The Lost World of the Kalahari, Laurens van der Post, Penguin, 1958

Macroeconomics, Lumsden, Attiyeh, and Bach, Prentice-Hall, 1970

Maximum Principles in Analytical Economics, Paul A. Samuelson, MIT, Cambridge, Massachusetts, 1970, https://www.nobelprize.org/uploads/2018/06/samuelson-lecture.pdf

Meadowland, John Lewis-Stempel, TransWorld Publishers, 2014

Microeconomics, Lumsden, Attiyeh, and Bach, Prentice-Hall, 1970

Money and Government, Robert Skidelsky, Penguin, 2019

The Modern British State, Halford John Mackinder, G. Philip/Forgotten Books, 1914

Numbers Guide, Richard Stutely, Economist Books, 1992

Nix Farm Management Pocketbook, The Anderson Centre, updated annually

The Paradox of Productivity: Agricultural Productivity Promotes Food System Inefficiency, Tim G Benton & Rob Bailey, Cambridge University Press, 2019, https://doi.org/10.1017/sus.2019.3

The Private Life of the Hare, John Lewis-Stempel, TransWorld Publishers, 2019

Revisiting Samuelson's Foundations of Economic Analysis, Roger E Backhouse, University of Birmingham, 2014, https://pure-oai.bham.ac.uk/ws/files/20456636/ Backhouse_Foundations_version_3_.pdf

The Running Hare: The Secret Life of Farmland, John Lewis-Stempel, TransWorld Publishers, 2016

Science, Adam Hart-Davies, Dorling Kindersley, 2012

The Secret Life of the Owl, John Lewis-Stempel, TransWorld Publishers, 2017

Serengeti Shall not Die, Grzimer, Hamish Hamilton, 1960

Shepherd's Life: A Tale of the Lake District, James Rebanks, Penguin, 2016

Silent Spring, Rachel Carson, Penguin Modern Classics, 2000

Six Inches of Soil: How to Heal Our Soils, Ourselves and Our Communities through Regenerative Farming, Molly Foster, Priya Kali, Jeremy Toynbee, 5m Books, 2024

The Small Farmer, H.J. Massingham, Collins, 1947

The UK Solar Energy Resource and the Impact of Climate Change, Dougal Burnett, Edward Barbour, Gareth P. Harrison, Elsevier, 2014, https://doi.org/10.1016/j.renene.2014.05.034

The Wild Life of the Fox, John Lewis-Stempel, TransWorld Publishers, 2020

The Wood: The Life & Times of Cockshutt Wood, John Lewis-Stempel, TransWorld Publishers, 2018

Wilding, Isabella Tree, Picador, 2018

INDEX

More from 5m Books

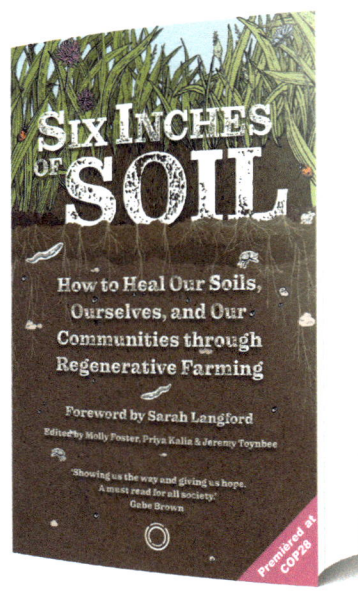

Six Inches of Soil

How to Heal Our Soils, Ourselves and Our Communities Through Regenerative Farming

Edited by Molly Foster, Priya Kalia, Jeremy Toynbee

Six Inches of Soil, the film and this companion book, is the inspiring story of three British farmers standing up to the industrial food system and transforming the way they produce food – to heal the soil, benefit our health and provide for local communities.

Apr 2024 | ISBN 9781917159005
£19.99 | $29.99 | €35.99 | 360p PB

9781739103927
Nov 2023
264p
PB
£14.99
$24.99
€17.99

A Dairy Story
Our Journey to Cow-with-Calf Dairy Farming

David Finlay, Wilma Finlay

A Dairy Story is the full, no holds barred story of their journey from conventional dairy farming to an organic, 100% pasture, regenerative, cow-with-calf dairy that's now at the forefront of a global movement to transform an industry.

9780578536729
Mar 2021
299p
PB
£16.99
$24.99
€17.99

For the Love of Soil
Strategies to Regenerate Our Food Production Systems

Nicole Masters

For the Love of Soil equips producers with knowledge, skills and insights to regenerate ecosystem health and grow farm/ranch profits.

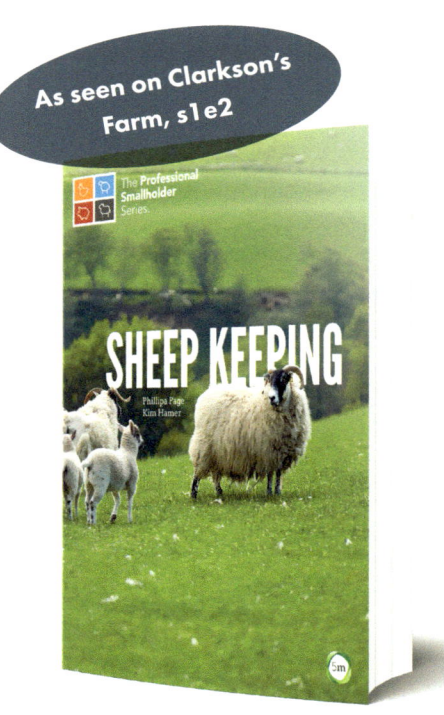

As seen on Clarkson's Farm, s1e2

The Professional Smallholder Series.

SHEEP KEEPING

Phillipa Page
Kim Hamer

Sheep Keeping

Phillipa Page and Kim Hamer

As a sheep keeper have you ever wanted to go into more depth with your vet and gain expert advice that will keep your flock healthy and happy? This guide provides a depth of veterinary information to complement local veterinary consultations. Advice is offered on everyday issues such as feeding and nutrition, housing, organ systems and their function, disease, treatments and sheep behaviour.

Sheep Keeping is a comprehensive handbook, providing veterinary relevant information to the sheep smallholder.

Nov 2017 | ISBN 9781910455937
£24.95 | $39.95 | €29.95 | 216p HB

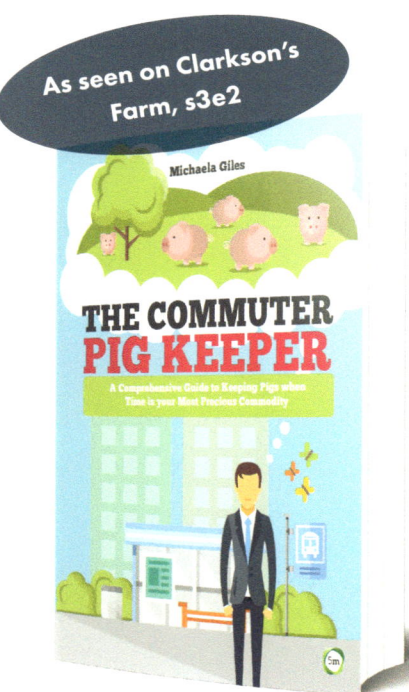

As seen on Clarkson's Farm, s3e2

Michaela Giles

THE COMMUTER PIG KEEPER

A Comprehensive Guide to Keeping Pigs when Time is your Most Precious Commodity

The Commuter Pig Keeper

A Comprehensive Guide to Keeping Pigs when Time is Your Most Precious Commodity

Michaela Giles

Aimed at people with busy schedules this instructive book gives practical information about how to manage a small herd and keep pigs happy and healthy under the time constraints of modern life.

The Commuter Pig Keeper is all-inclusive covering various breeds both as breeding herds and food sources. Topics addressed include pig rearing, breeding, housing, and handling techniques.

Aug 2016 | ISBN 9781910455531
£24.95 | $39.95 | €29.95 | 288p PB

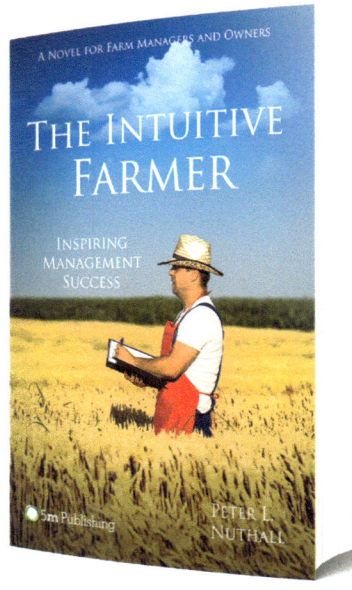

The Intuitive Farmer

Inspiring Management Success

Peter Nuthall

The Intuitive Farmer follows on from successful business management books such as *The Goal*, which communicate business ideas and strategies in novel form. This is the first such book applied to agricultural management practices, providing a dependable source for farmers, agricultural and farm management students and people involved in agriculture industries. By the end of the novel the reader will have absorbed important farm management principles and practices through the activities and findings of the group.

Jan 2016 | ISBN 9781910455135
£19.90 | $29.90 | €23.90 | 232p PB

The Veterinary Book for Dairy Farmers

Roger Blowey

The ever-changing world of cattle farming requires farmers to be up-to-date with best-practice procedures and the latest advances in husbandry techniques. Now in its 4th edition Roger Blowey's updated version of the acclaimed *A Veterinary Book for Dairy Farmers* deals with newly emerging problems in cattle farming as well as covering the necessary knowledge required for maintenance and prosperity.

Oct 2016 | ISBN 9781908397775
£55 | $85 | €66 | 552p HB

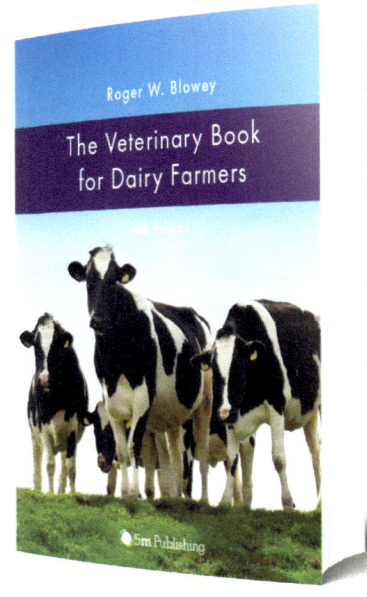

9780955501159
Oct 2013
600p
HB
£115.00
$179.95
€140.00

Managing Pig Health 2nd Edition
A Reference for the Farm

Edited by John Carr

This new updated edition of *Managing Pig Health* offers a fresh and comprehensive guide to practical veterinary information for pig farmers, veterinarians and technologists around the world.

9781789181197
Jun 2021
504p
HB
£55.00
$85.00
€66.00

The Veterinary Book for Beef Farmers

Lee-Anne Oliver and Roger Scott

A comprehensive forage-to-fork book on beef farming that delivers veterinary level information to farmers and farm-related information to vets.

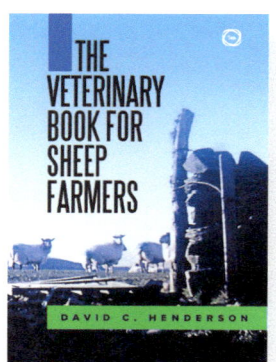

978190336630
Dec 2002
738p
HB
£24.95
$39.95
€29.95

The Veterinary Book for Sheep Farmers

David Henderson

This is a practical manual focusing principally on the needs of sheep farmers and shepherds in the United Kingdom but covering material applicable to sheep-farming situations worldwide.

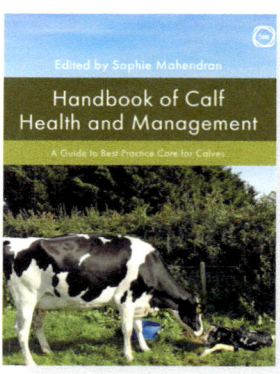

9781789181340
Oct 2021
360p
HB
£55.00
$85.00
€66.00

Handbook of Calf Health and Management
A Guide to Best Practice Care for Calves

Edited by Sophie Mahendran

A guide to calf management authored by vets who also have their own herd. The focus is on rearing an animal with optimal health for enhanced productivity with a focus on welfare and best practice.